4th International Conference on Biomolecular Engineering (ICBE 2013)

Fort Lauderdale, Florida, USA
13 – 16 January 2013

ISBN: 978-1-5108-1596-4

Printed from e-media with permission by:

Curran Associates, Inc.
57 Morehouse Lane
Red Hook, NY 12571

Some format issues inherent in the e-media version may also appear in this print version.

Printed by Curran Associates, Inc. (2015)

For permission requests, please contact AIChE
at the address below.

AIChE
120 Wall Street, FL 23
New York, NY 10005-4020

Phone: (800) 242-4363
Fax: (203) 775-5177

www.aiche.org

Additional copies of this publication are available from:

Curran Associates, Inc.
57 Morehouse Lane
Red Hook, NY 12571 USA
Phone: 845-758-0400
Fax: 845-758-2634
Email: curran@proceedings.com
Web: www.proceedings.com

TABLE OF CONTENTS

Modeling

Jin Liu
Ryan Bradley
Ramakrishnan Natesan
Peter Huwe
Ravi Radhakrishnan

Experiments

David Eckmann
Vladimir Muzykantov
Tatyana Svitkina
Paul Janmey
Tobias Baumgart
Wei Guo
Mark Lemmon

Model-Based Design and Optimization of Functionalized Nanocarriers for Vascular Targeted Drug Delivery

Modeling

Jin Liu
Ryan Bradley
Ramakrishnan Natesan
Peter Huwe
Ravi Radhakrishnan

Experiments

David Eckmann
Vladimir Muzykantov
Tatyana Svitkina
Paul Janmey
Tobias Baumgart
Wei Guo
Mark Lemmon

Physical and Biochemical Modeling of Proteins, Membranes, and Signal Transduction

Multiscale modeling enables spanning range of time and length scales

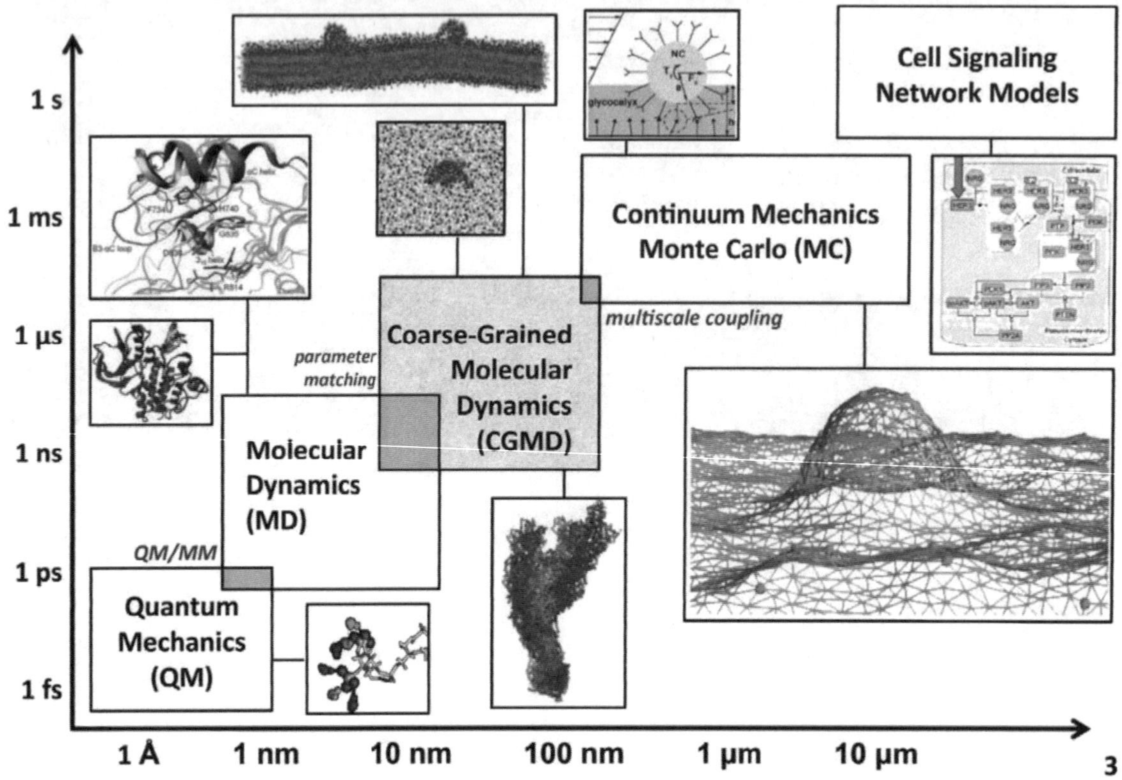

3

Outline

- Top down Modeling: Design of functionalized nanocarriers for binding efficiency, selectivity
- Bottom-up Modeling: Multiscale modeling platform for biophysical models of intracellular trafficking
- Bottom-up Modeling: Molecular modeling of clinical mutations in oncogenic kinases

University of Pennsylvania Department of Bioengineering

Targeted Drug Delivery using Functionalized Nanocarriers

Peer *et al.*, Nature Nanotechnology, 2007

Specific receptor proteins expressed on pathological cells provide good target. Coating specific ligands onto the nanocarrier surface helps enhance the efficiency of binding.

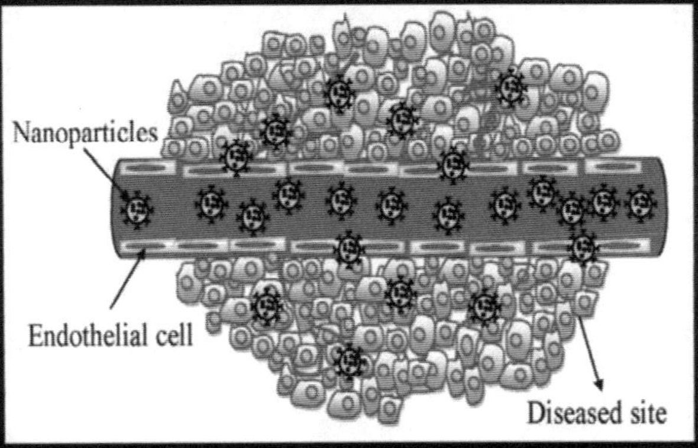

Time scales:
1. Hydrodynamic
2. Viscous/Brownian relaxation
3. Particulate Collision
4. Adhesion

Length scales:
1. Hydrodynamic (macro)
2. Near-wall (meso)
3. Molecular recognition (nano)

How do the NC size, shape, ligand surface coverage etc. affect efficiency of targeting in vitro and in vivo?

University of Pennsylvania SEAS/Perelman SOM

Minimal Model for Nanocarrier Adhesion

Objective: How do we quantify NC binding avidity to cells mediated by antibody-antigen interactions?

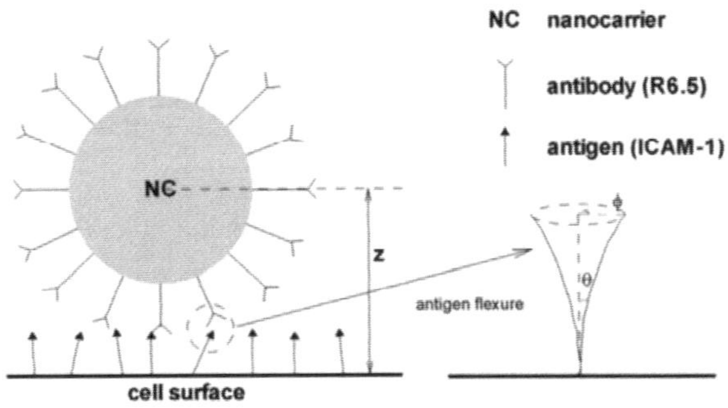

Liu *et al.* (2010) *PNAS* **107** 16530-16535

- Spherical 100nm NCs coated with uniformly distributed N_{ab} antibodies (Ab). The saturation Ab surface coverage (100%) σ_s=220Ab/NC in experiments.
- Antibody-antigen interaction is treated using the Bell model:

$$\Delta G_r(d) = \Delta G_0 + \frac{1}{2}k_0 d^2 \qquad k_0 : \text{bond force constant}$$

- Antigen flexure accounted for by orientational-bias MC sampling of θ and ϕ

University of Pennsylvania SEAS/ Perelman SOM

Avidity: Absolute Free Energy of Binding of NC to Cells

Potential of Mean Force (PMF)

The binding avidity or association constant K_a directly measures the binding efficiency, for reaction

L+R ⇔ LR,

$$K_a = \frac{[LR]}{[L][R]} = \frac{1}{[L]} \times \frac{p_1}{p_0}$$

bound

unbound

Liu *et al.* (2010) *PNAS* **107** 16530-16535

The final form for K_a is:

PMF W(z)

$$K_a = \frac{(N_{ab}/N_b)\Delta\omega}{8\pi^2} \times \frac{A^{(1)}_{R,b} \times A^{(2)}_{R,b} \times \ldots \times A^{(N_b)}_{R,b}}{A^{(1)}_{R,ub} \times A^{(2)}_{R,ub} \times \ldots \times A^{(N_b)}_{R,ub}} \times A_{NC,b} \int e^{-\beta W(z)} dz$$

NC rotational entropy change

antigen entropy change

NC translational entropy change

Super-resolution florescence

AFM

Directly measurable

Potential of Mean Force (σ_s=75%, 162Abs)

PMF profile

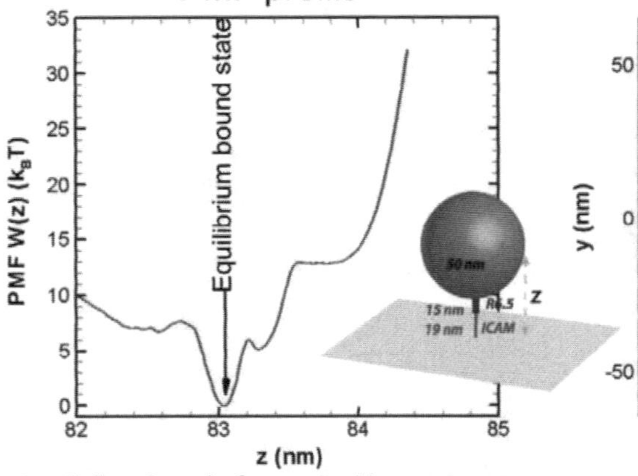

bond distribution

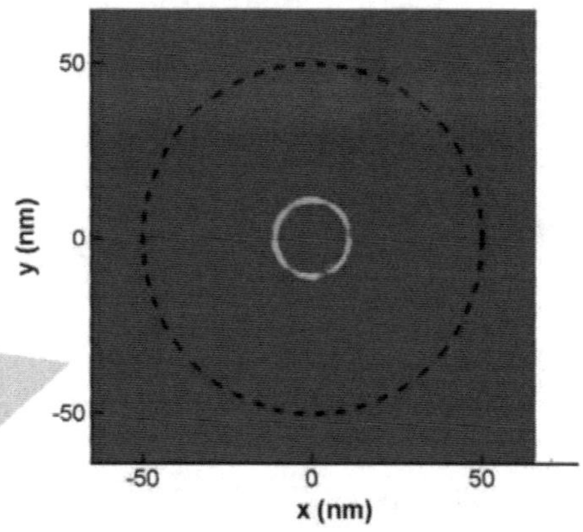

- 3 firm bonds formed with an energy well ~32k_BT

$$K_a = \frac{(N_{ab}/3)\Delta\omega}{8\pi^2} \times \frac{\pi(r_o^2 - r_i^2)(r_o - r_i)^4 \int e^{-\beta W(z)} dz}{(\pi r_o^2)^2} = 5.9 \times 10^{10}\,\text{nm}^3$$

Liu *et al.* (2010) *PNAS* **107** 16530-16535

- Annulus distribution with outer radius r_o=12.5nm and inner radius r_i=9.7nm
- The areas can be calculated as:

$$A^{(1)}_{R,b} = \pi r_o^2 \qquad A^{(2)}_{R,b} = \pi\left(r_o^2 - r_i^2\right)$$

$$A^{(3)}_{R,b} = A_{NC,b} = \left(r_o - r_i\right)^2$$

$$A^{(1)}_{R,ub} = A^{(2)}_{R,ub} = A^{(3)}_{R,ub} = \pi r_o^2$$

NC Dissociation Constant, K_d

Model prediction

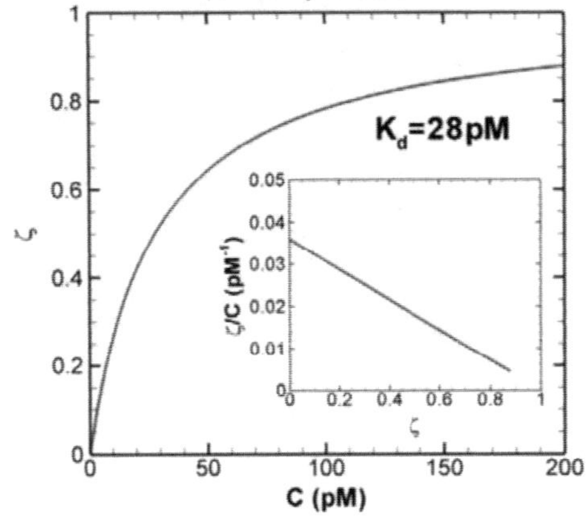

In vitro experiment

Muro *et al.* J. Pharmacol. Exp. Ther. 317 1161 (2006)

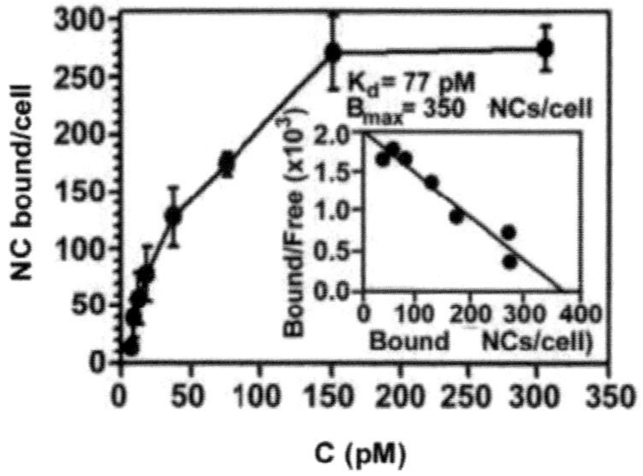

According to Langmuir binding model, the fractional binding:

$$\zeta = \frac{C/K_d}{1+C/K_d} \qquad K_d = 1/K_a$$

Liu *et al.* (2010) *PNAS* **107** 16530-16535

K_d	Model (27°C)	Experiment (4°C)
R6.5-ICAM-1	3.2nM	8.5nM
NC-cell	28pM	77pM

Zero-fit model predictions are consistent with in vitro cellular experiments

Effect of Nanocarrier Surface Coverage σ_s (# antibodies per NC)

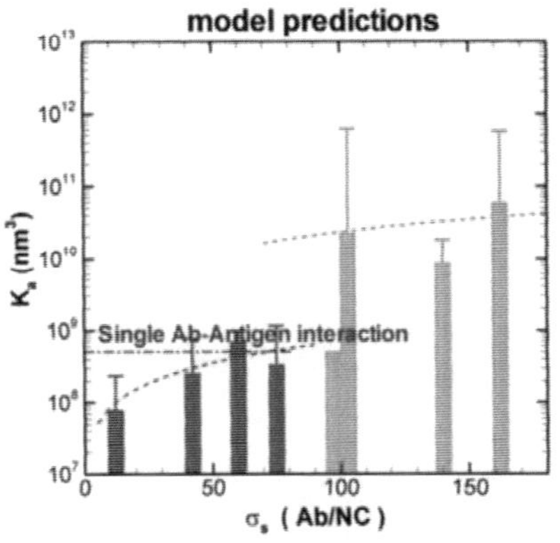

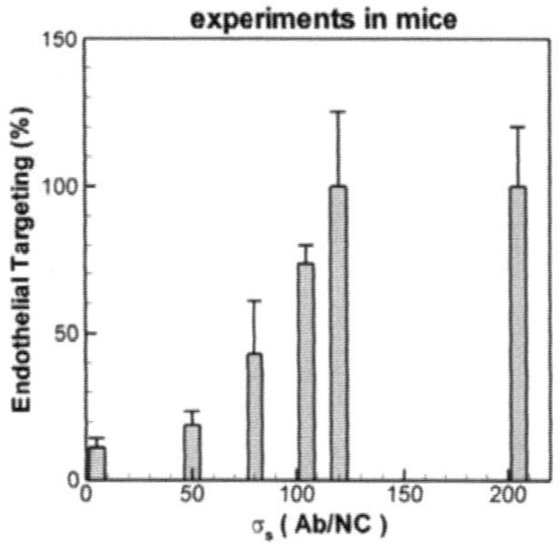

- A threshold at σ_s ~45% (100Ab/NC), the binding affinity abruptly drops below that of single antibody to antigen

Liu *et al.* (2010) *PNAS* **107** 16530-16535

$$K_a = \frac{(N_{ab}/N_b)\Delta\omega}{8\pi^2} \times \frac{A_{R,b}^{(1)} \times A_{R,b}^{(2)} \times \ldots \times A_{R,b}^{(N_b)}}{A_{R,ub}^{(1)} \times A_{R,ub}^{(2)} \times \ldots \times A_{R,ub}^{(N_b)}} \times A_{NC,b} \int e^{-\beta W(z)} dz$$

- Linear dependence below and above the threshold at fixed multivalency, dotted lines.
- Exponential reduction because of the multivalency change (from 3 to 2) around σ_s ~45%

Model predictions are consistent with results of in vivo mice experiments

NC Rupture Force (AFM Experiment)

The probability distribution of rupture force:

$$\begin{cases} P(f) = k^0 \exp\left(\dfrac{\gamma f}{k_B T}\right) \exp\left\{\dfrac{k^0 k_B T}{\gamma r_f}\left[1-\exp\left(\dfrac{\gamma f}{k_B T}\right)\right]\right\} \\ k^0 = \dfrac{k_B T}{h}\exp\left(\dfrac{-\Delta G}{k_B T}\right) \qquad \gamma \sim 0.2\text{nm} \\ r_f : \text{Loading rate between 100000 and 200000pN/s} \end{cases}$$

Model predicts: 230±41pN

AFM experiments (89 trials): 316±48pN

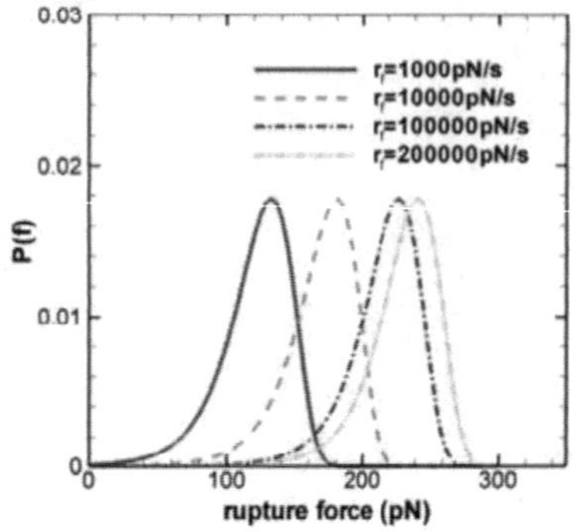

AFM set-up

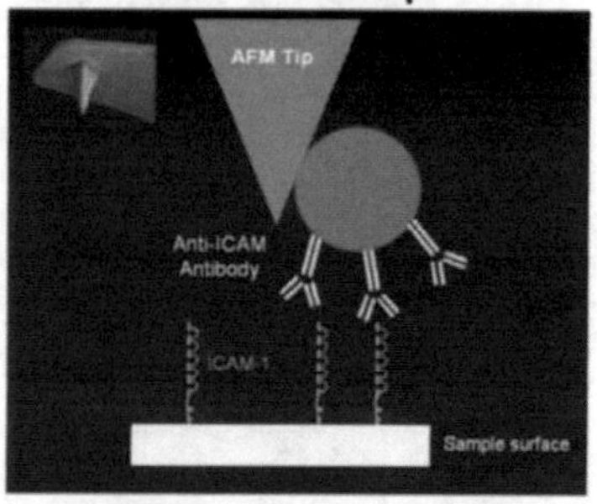

AFM measurement

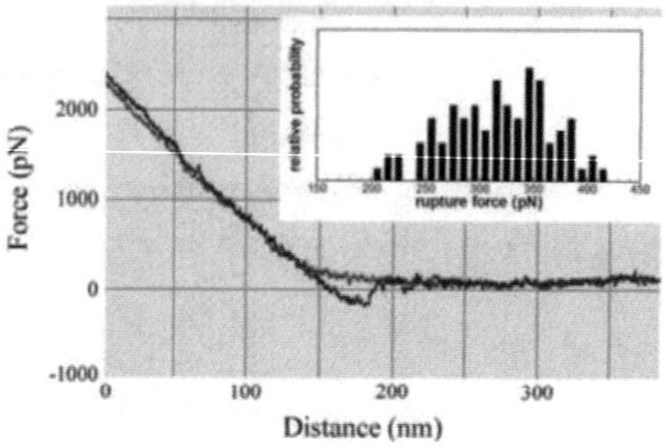

Model prediction of rupture force distribution is consistent with AFM experiment

Effect of Flow on NC Adhesion: Shear Enhanced Binding and Rolling Behavior

J. Liu, et al., Biophys J, 2011

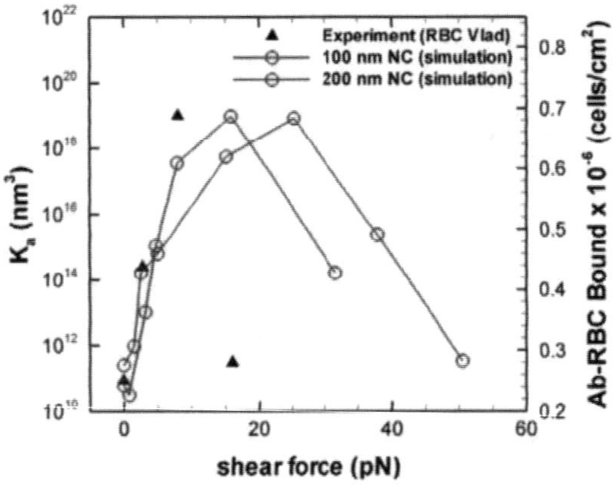

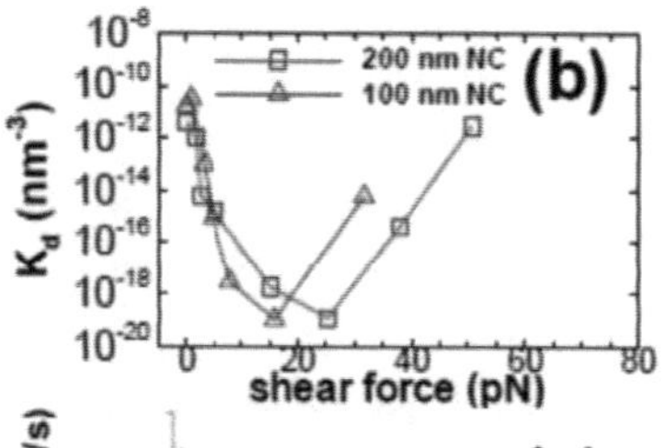

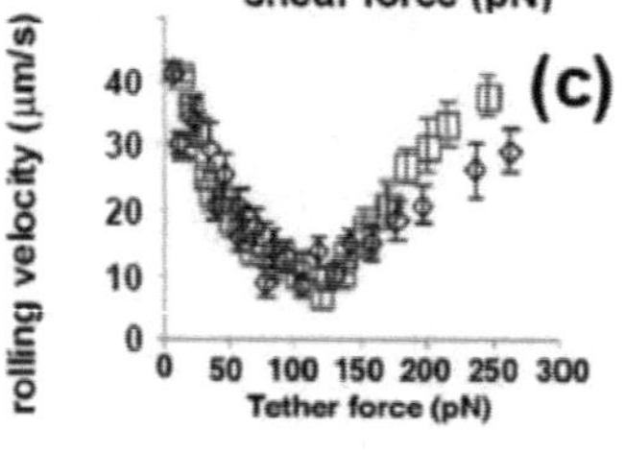

Intriguingly, model achieves simultaneous thermodynamic, physiological, and mechanical consistency. This holds potential for such minimal models to guide the design of nanocarriers for targeted drug delivery

Conclusions

- Top Down Physically Based Models are minimal, but offer a powerful approach to interpret mesoscale experiments and subject them to structural, mechanical, and energetic consistency
- Independent comparisons to several experiments enable direct parameter estimation. Hence predictive models can be constructed with zero fitted parameters
- In the event that certain parameters are unavailable from experiments, once can (in principle) resort to molecular models for its determination
- This approach is quantitative and enables to interpret biology on the grounds of biophysical chemistry; it is also predictive and can be useful for rational design

University of Pennsylvania Department of Bioengineering

Intracellular Targeting

- A predictive modeling platform for protein-induced curvature of membranes has potential to significantly transform/interpret mechanisms intracellular trafficking in targeted drug delivery

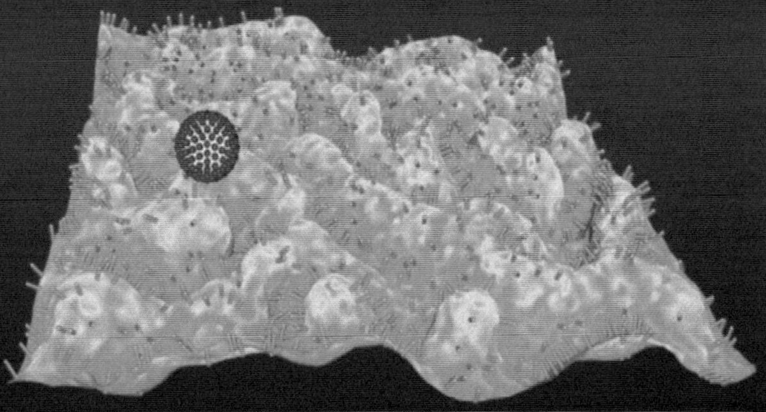

University of Pennsylvania Department of Bioengineering

Outline

- Top down Modeling: Design of functionalized nanocarriers for binding efficiency, selectivity
- Bottom-up Modeling: Multiscale modeling platform for biophysical models of intracellular trafficking
- Bottom-up Modeling: Molecular modeling of clinical mutations in oncogenic kinases

University of Pennsylvania — Department of Bioengineering

Intracellular Trafficking and Membrane Curvature Remodeling

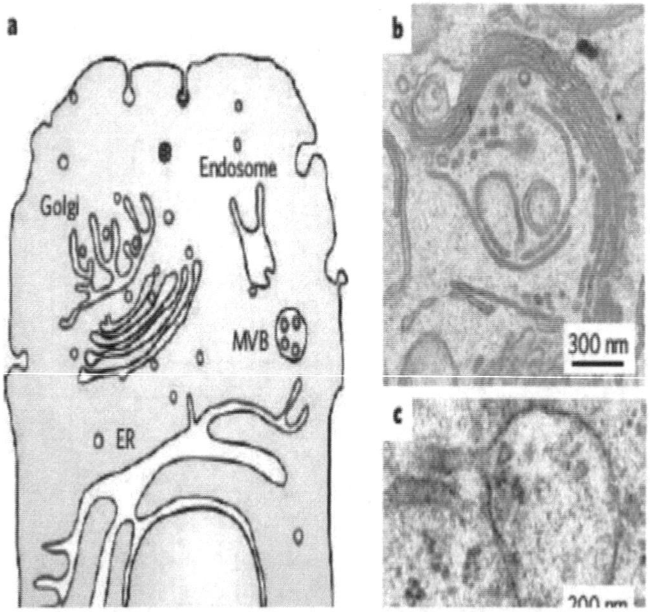

McMahon, Galop, 2005

Triangulated Surface Model for Membranes

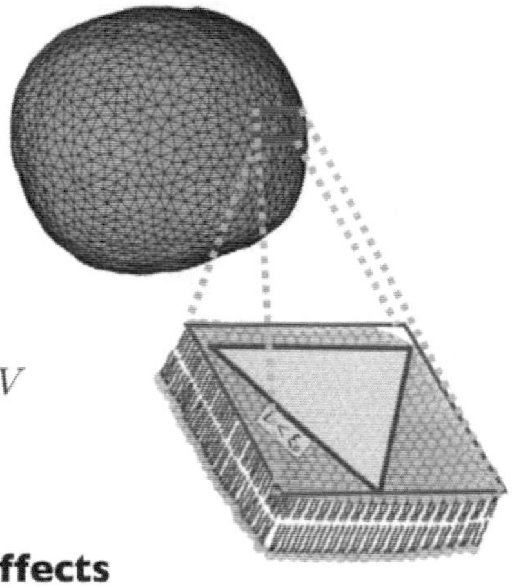

Isotropic Effects

$$\mathscr{H}_{\text{sur}} = \kappa \sum_{v=1}^{N} \left(\frac{c_1(v) + c_2(v)}{2} - C_0 \right)^2 A(v) - \Delta p\, V$$

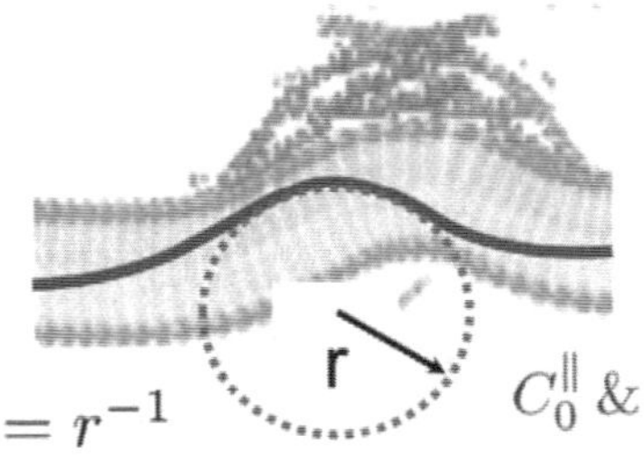

Anisotropic Effects

$\kappa_{\|}$ & $\kappa_{\perp}$ & $\kappa_{\times}$ → directional bending stiffness

$C_0^{\|} = r^{-1}$ $\quad$ $C_0^{\|}$ & $C_0^{\perp}$ & $C_0^{\times}$ → directional spontaneous curvatures

$$\mathscr{H}_{\text{anis}} = \sum_{v=1}^{N} A(v) \left\{ \frac{\kappa_{\|}}{2} \left(H_{\|}(v) - c_0^{\|} \right)^2 + \frac{\kappa_{\perp}}{2} \left(H_{\perp}(v) - c_0^{\perp}]^2 \right) \right\}$$

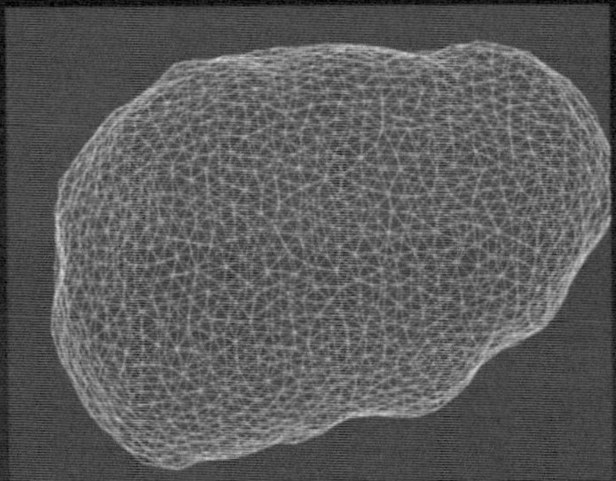

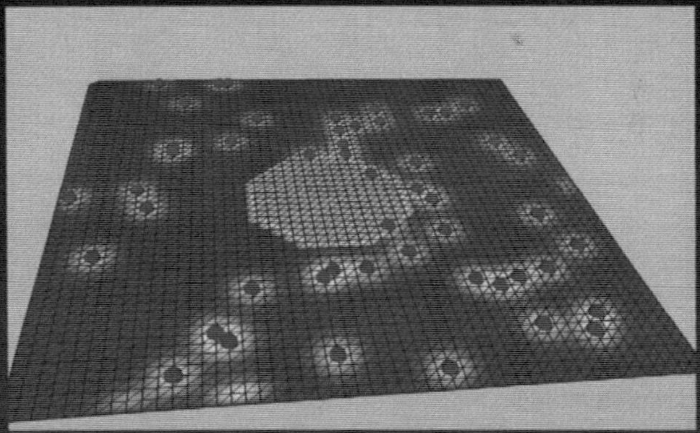

Vesicle/cell dynamics $\quad$ Membrane with isotropic spontaneous curvature fields

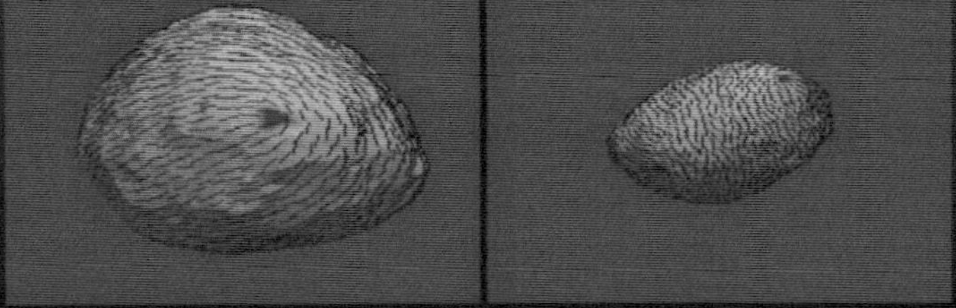

Membrane with anisotropic spontaneous curvature fields

University of Pennsylvania $\quad$ Department of Bioengineering

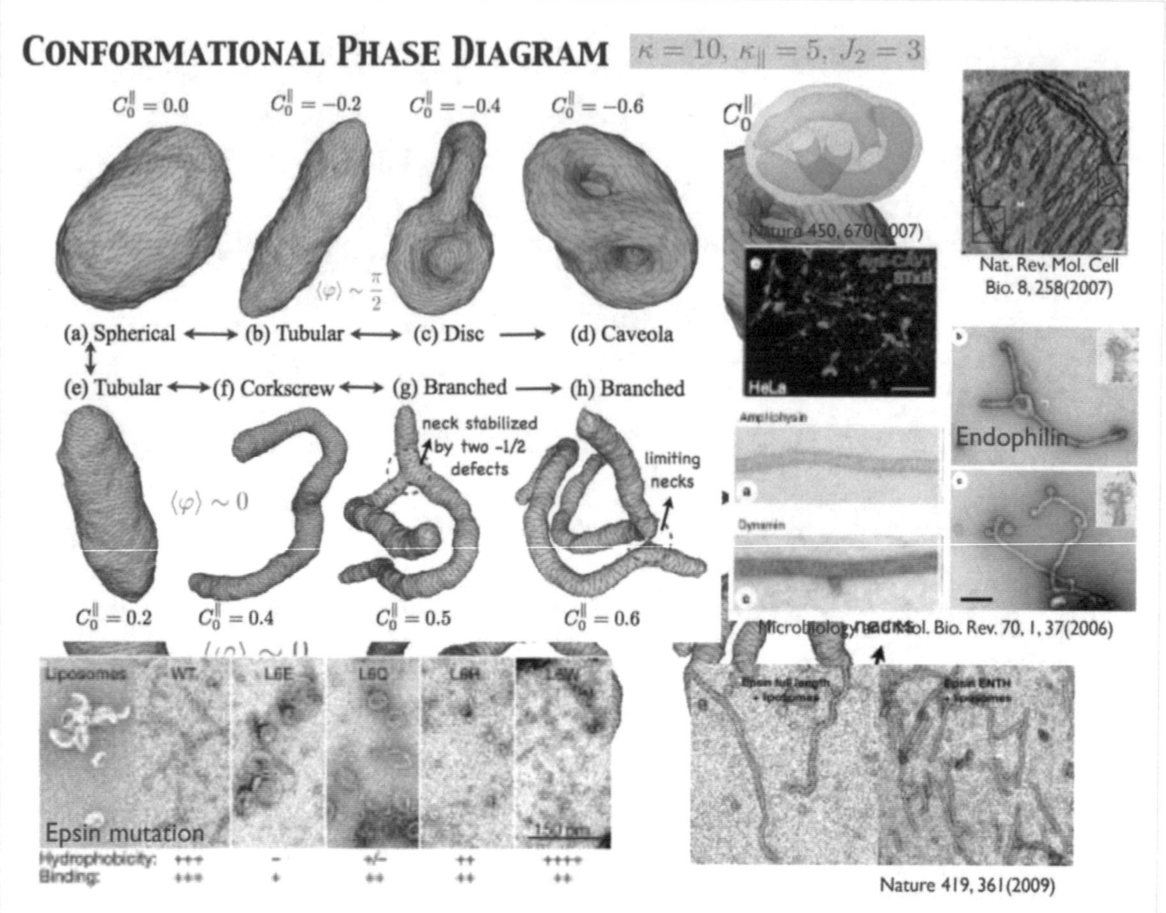
CONFORMATIONAL PHASE DIAGRAM
$\kappa = 10, \kappa_{\parallel} = 5, J_2 = 3$
$C_0^{\parallel} = 0.0$
$C_0^{\parallel} = -0.2$
$C_0^{\parallel} = -0.4$
$C_0^{\parallel} = -0.6$
$\langle\varphi\rangle \sim \frac{\pi}{2}$
(a) Spherical
(b) Tubular
(c) Disc
(d) Caveola
(e) Tubular
(f) Corkscrew
(g) Branched
(h) Branched
neck stabilized by two -1/2 defects
limiting necks
$\langle\varphi\rangle \sim 0$
$C_0^{\parallel} = 0.2$
$C_0^{\parallel} = 0.4$
$C_0^{\parallel} = 0.5$
$C_0^{\parallel} = 0.6$
Nature 450, 670(2007)
HeLa
Nat. Rev. Mol. Cell Bio. 8, 258(2007)
Endophilin
Microbiology and Mol. Bio. Rev. 70, 1, 37(2006)
necks
Liposomes
WT
L6E
L6Q
L6H
L6W
Epsin mutation
150 nm
Hydrophobicity: +++ - +/- ++ ++++
Binding: +++ + ++ ++ ++
Nature 419, 361(2009)

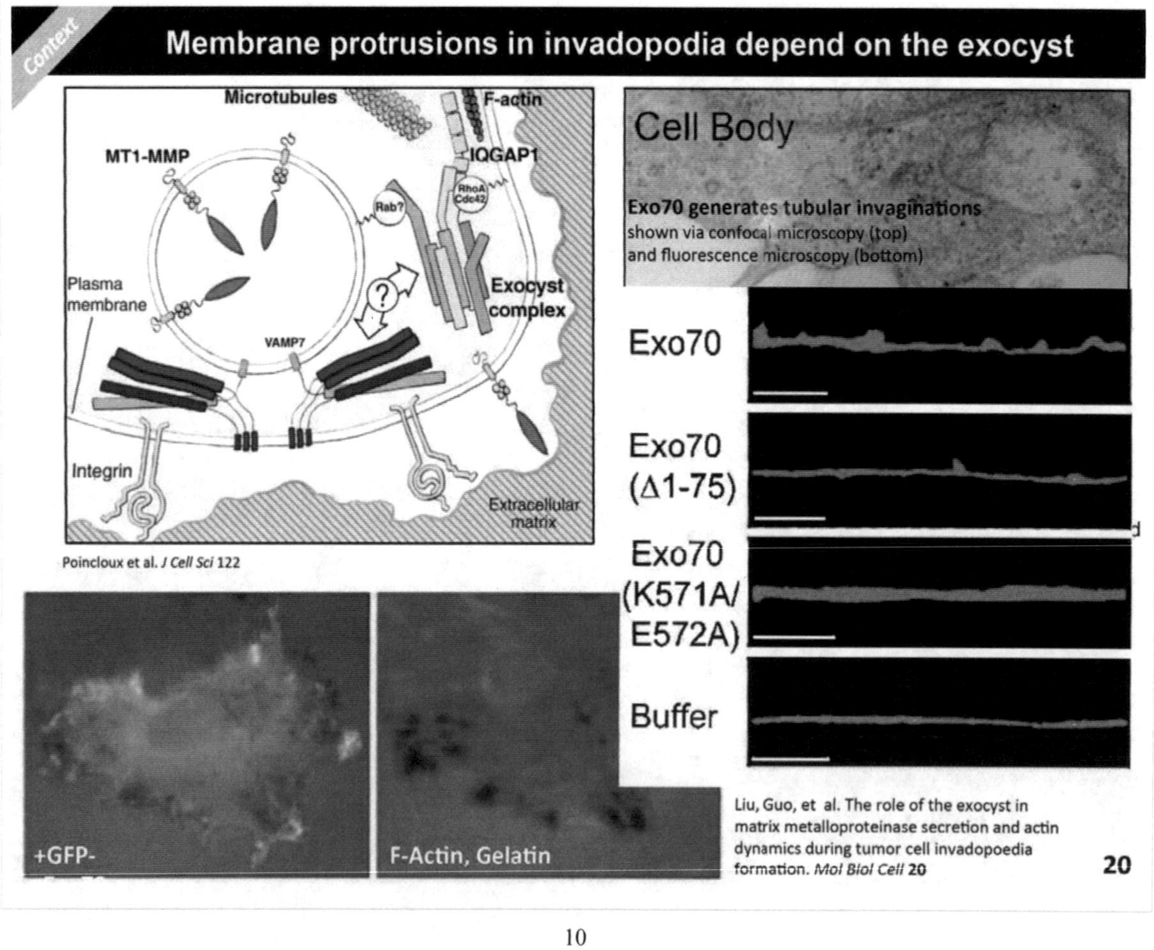
Context
Membrane protrusions in invadopodia depend on the exocyst
Microtubules
F-actin
MT1-MMP
IQGAP1
Rab?
RhoA Cdc42
Plasma membrane
Exocyst complex
VAMP7
Integrin
Extracellular matrix
Poincloux et al. J Cell Sci 122
Cell Body
Exo70 generates tubular invaginations
shown via confocal microscopy (top)
and fluorescence microscopy (bottom)
Exo70
Exo70 (Δ1-75)
Exo70 (K571A/ E572A)
Buffer
+GFP-
F-Actin, Gelatin
Liu, Guo, et al. The role of the exocyst in matrix metalloproteinase secretion and actin dynamics during tumor cell invadopoedia formation. Mol Biol Cell 20
20

Model

Coarse-grained Exo70 model from crystal, helix alignment

All-atom Exo70 monomer
Gaps filled by MODELLER, then relaxed using 20ns simulations in CHARMM27/GROMACS

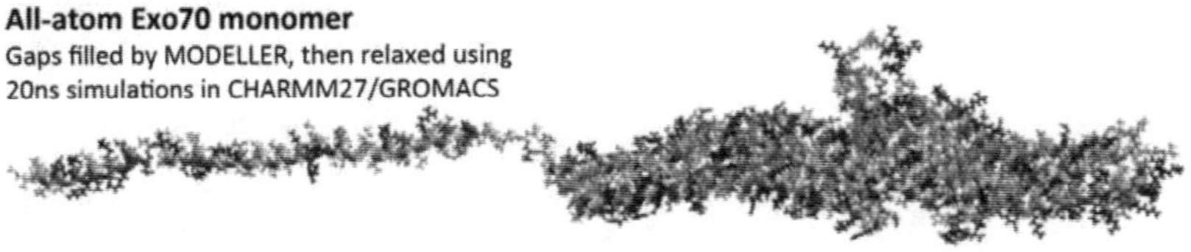

CGMD dimer model
Assisted in clasping the helix, then removed

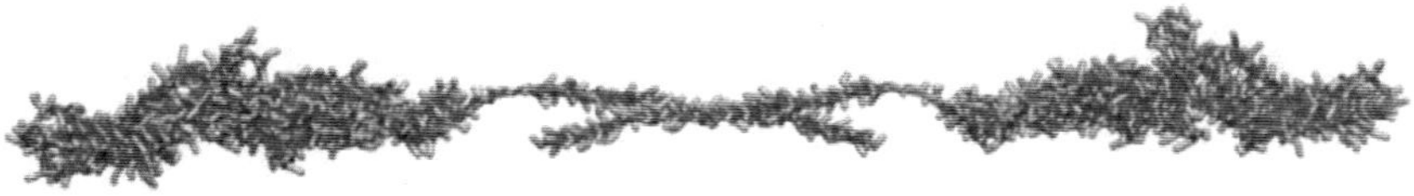

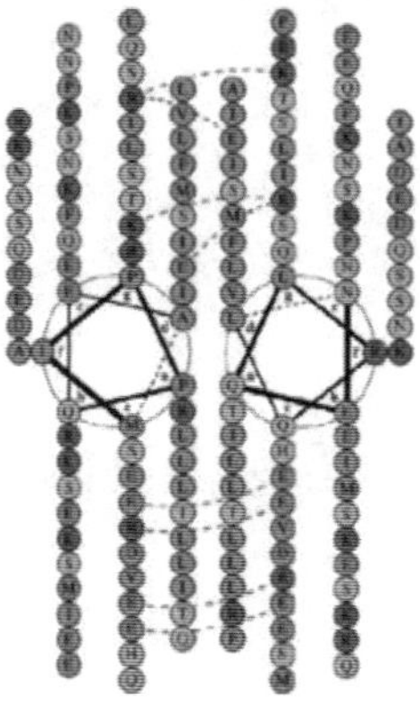

Dimer α-Helices aligned
By manually matching charged residues

CGMD dimer model on membrane
Placed, minimized, equilibrated

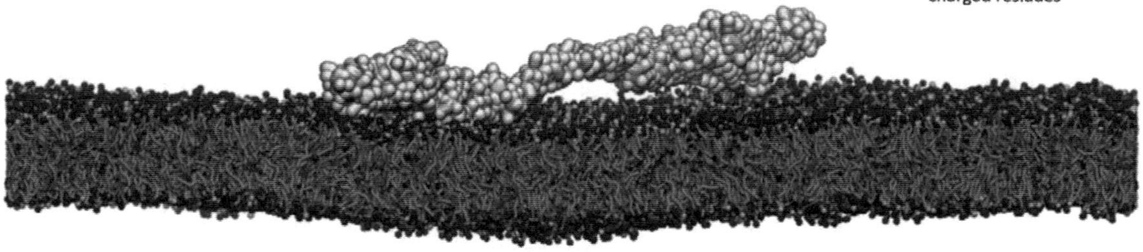

21

Analysis

Membrane undulations couple to the curvature field

Height map (red negative, blue positive) indicate that the induced curvature field is modulated by membrane undulations, and depends on the proximity to other proteins in the periodic images.

Anti-parallel dimer simulations
MARTINI Force Field, GROMACS integrator, Standard MARTINI parameters
12,800 lipids, 4:1 DOPC/DOPS
40 fs timestep, 150 ns trajectory

22

Molecular mechanism of negative curvature suggested by simulations

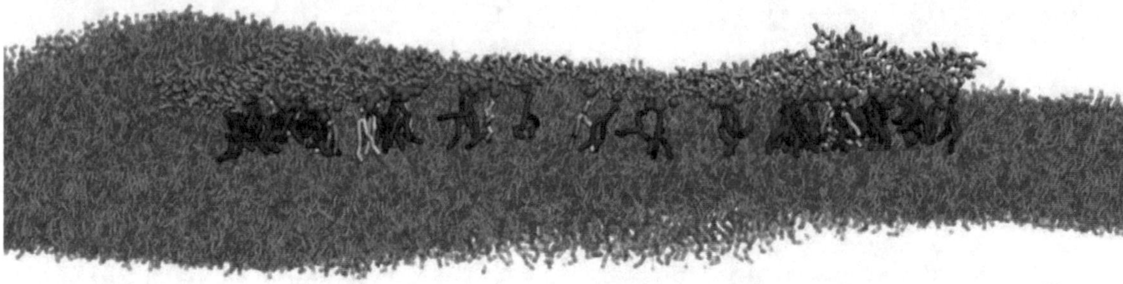

Exo70 may pattern negative charge
Negatively charged lipids may cluster around positive charges on Exo70, inducing curvature (via area asymmetry) by compressing these lipids

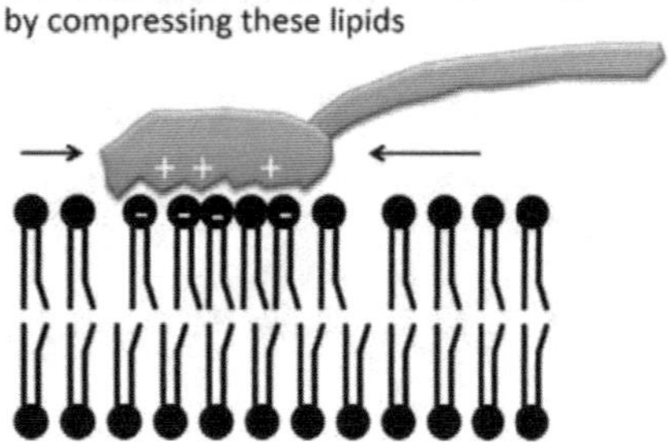

Membrane-associating residues
make persistent contacts with lipids

404	ASP	407	LYS
408	GLU	336	ARG
329	GLU	330	ASN
343	ASP	484	VAL
501	Lys	502	Glu

Exo70 linkers scaffold monomers
at distances necessary for negative curvature, and may also directly apply force to the membrane via monomer contacts

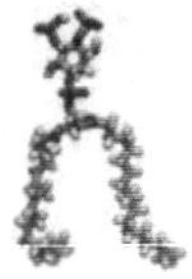

Future work

Addition of PIP2
will further resolve the role of binding to the strongly-charged PIP2 lipids, and further analysis will quantify the effect of charge on the stress map

23

Mesoscale model links MD simulations to experiment

Threshold for tubulation
This model requires a threshold radius of curvature R<R* to induce membrane invaginations.

This threshold links the microscopic MD simulations – which report a_0 – to experiments in which the binding and deletion mutants abolish membrane protrusions.

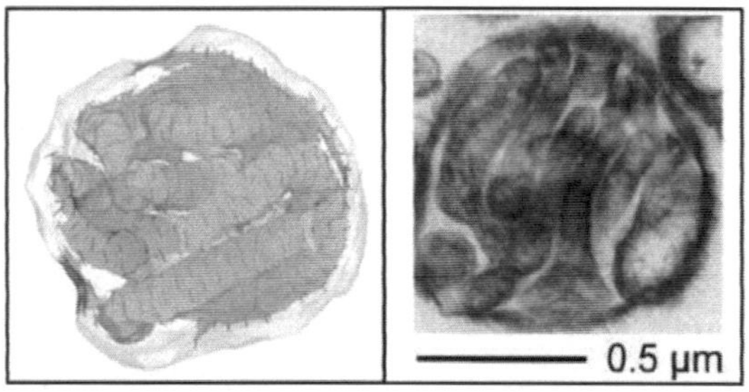

continuum mechanics simulation

confocal fluorescence microscopy

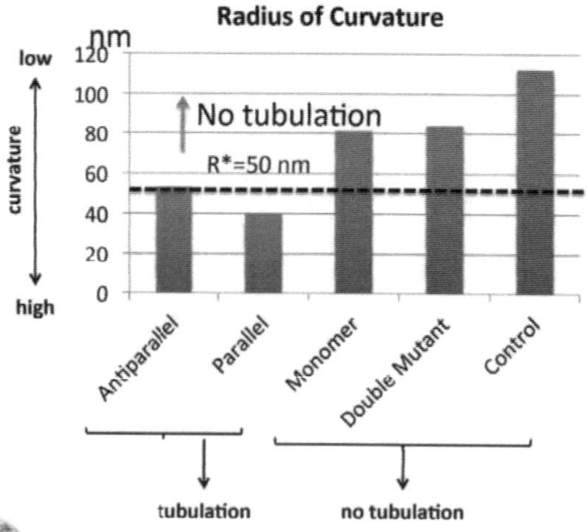

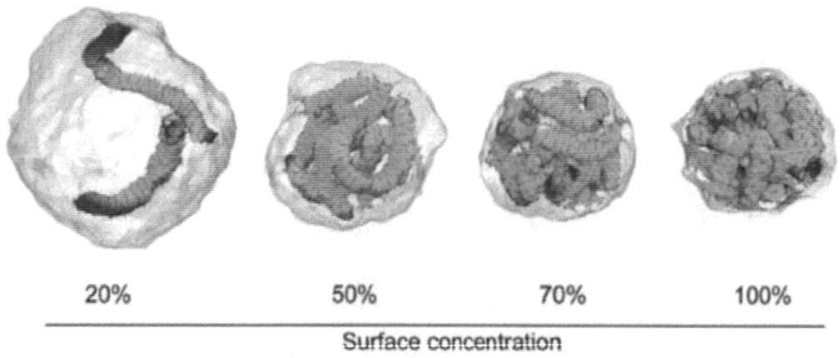

Higher surface concentrations of Exo70 result in greater membrane invaginations

24

Conclusions

- In situations where molecular mechanisms of membrane morphologies are not known, we can combine molecular simulations together with continuum simulations to suggest possible mechanisms and link to cellular/mutagenesis data quantitatively

Outline

- Top down Modeling: Design of functionalized nanocarriers for binding efficiency, selectivity
- Bottom-up Modeling: Multiscale modeling platform for biophysical models of intracellular trafficking
- Bottom-up Modeling: Molecular modeling of clinical mutations in oncogenic kinases

University of Pennsylvania   Department of Bioengineering

Insilico Oncology: Tumor Complexity & Onset of Fragility at Multiple Scales

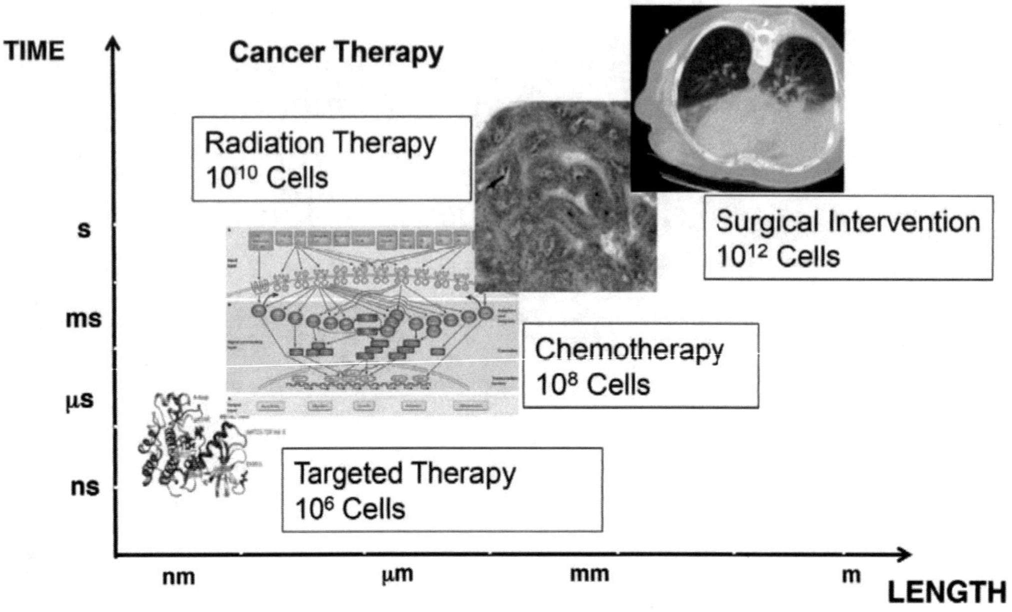

Receptor Tyrosine Kinases (RTKs)

Lemmon, M.A. and Schlessinger, J. *Cell* (2010)

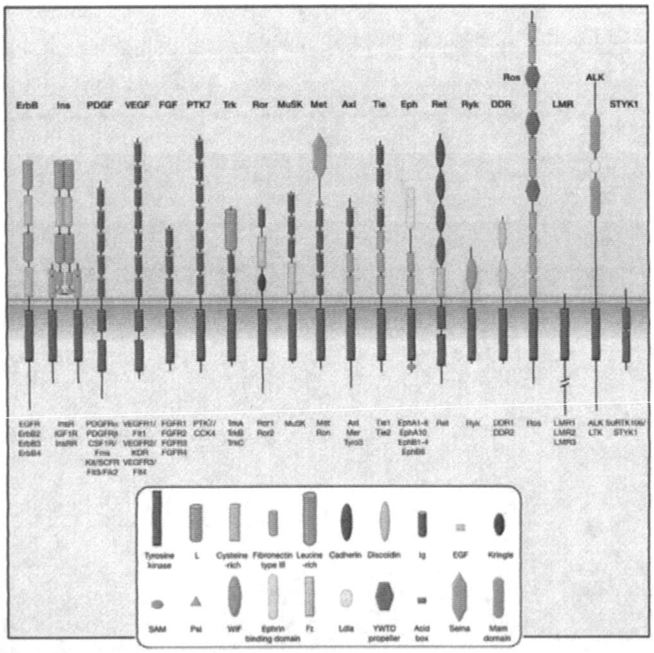

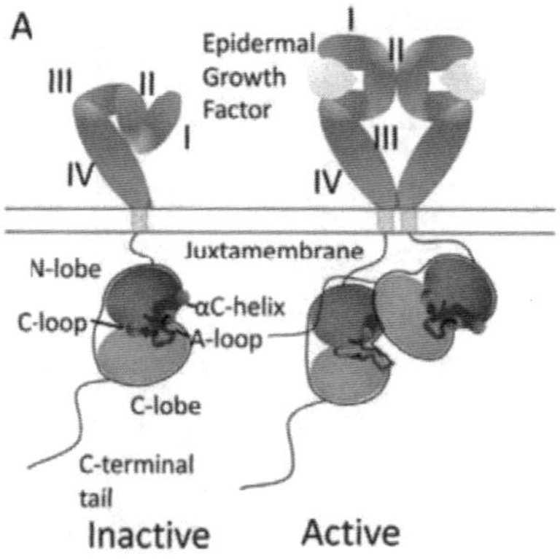

Molecular Models-
Mutational Landscape
Functional Effects of
Kinase Activation

University of Pennsylvania Department of Bioengineering

Currently Known Activating Kinase Mutations by Subdomain

	Lung	Colon	Skin	Breast	Prostate	Leukemia
EGFR	PL αC αCβ4 HC AD1	–	αCβ4	αC	αC	–
ErbB2	αC αCβ4	αCβ4	–	αCβ4 αC	–	–
ErbB4	–	AL	AL	AL	–	–
PDGFRα	JM CT	–	–	–	–	AL
CSF1R/ Fms	–	–	–	–	–	CT
Kit/SCFR	JM	–	JM αC AL	–	–	JM AL
Flt3/Flk2	–	–	–	–	–	JM AL
VEGFR2/ KDR	CT		–	TM AL	–	–
FGFR1	AL	–	–	–	–	–
FGFR2	–	–	JM* αC* AL*	–	–	–
FGFR3	–	–	AL	–	–	AL
FGFR4	AL	–	–	–	–	–
Met	JM	αC	–	–	–	–
EphA1-8, 10	AL CT	–	–	–	–	–
LTK	AL CT	–	–	–	–	–

	Ovary	Kidney	Thyroid	Gastro Intestinal	Neuro Blastoma
EGFR	–	αC	αC AL AD1	–	–
ErbB2	αCβ4	–	–	–	–
ErbB4	–	–	–	αC	–
PDGFRα	–	–	–	AL JM	–
Kit/SCFR	AL	–	–	JM	–
Met	–	–	JM	–	–
Ret	–	–	AD2 JM AL	–	–
ALK	–	–	–	–	AL αC

JM: juxtamembrane domain
PL:P-loop
αC: αC-helix
αCβ4: αC-β4 loop
HC: Hydrophobic Core
AL: A-loop
AD1: Asymmetric Dimer Interface in ErbB family
AD2: Asymmetric Dimer Interface in Ret
CT: C-terminal Tail.

Shih et al, Cancers, 2011

RTKs Inhibitors Approved and in Use as Cancer Therapeutics

Name	Target	Company	Class
Bevacizumab (Avastin)	VEGF	Genentech	Monoclonal antibody
Cetuximab (Erbitux)	EGFR	Imclone/Bristol-Meyers Squib	Monoclonal antibody
Panitumumab (Vectibix)	EGFR	Amgen	Monoclonal antibody
Ranibizumab (Lucentis)	VEGF	Genentech	Monoclonal antibody
Trastuzumab (Herceptin)	Erb2	Genentech	Monoclonal antibody
Pegaptanib (Macugen)	VEGF	OSI/Pfizer	RNA Aptamer
Dasatinib (Sprycel)	Src/Bcr-Abl	Bristol-Meyers Squib	Small molecule
Erlotinib (Tarceva)	EGFR	Genentech/OSI	Small molecule
Gefitinib (Iressa)	EGFR	AstraZeneca	Small molecule
Imatinib (Gleevec)	Bcr-Abl	Novartis	Small molecule
Lapatinib (Tykerb)	EGFR/Erb2	GSK	Small molecule
Nilotinib (Tasigna)	Bcr-Abl	Novartis	Small molecule
Pazopanib (Votrient)	VEGFR1/2/3 PDGFR/c-kit	GlaxoSmithKline	Small molecule
Sorafenib (Nexavar)	RAF/VEGFR2/PDGFRB	Onyx/Bayer	Small molecule
Sunitinib (Sutent)	VEGFR2/PDGFRB c-kit/FLT3	Pfizer	Small molecule

Understanding how kinases activate would better our control of RTKs and help design better cancer therapeutics.

Shih et al, Cancers, 2011

Pediatric Neuroblastoma

- Arises from neural crest tissue along the sympathetic chain ganglion
- Very different phenotypes
- By far the most common cancer in infants
- Sometimes familial, mostly sporadic
- 5 year survival rate for high risk patients is 20 - 40%
- Anaplastic Lymphoma Kinase (ALK) is an RTK which plays a role in the development of neural crest
- In cancers, ALK is often mutated, over expressed, or fused to another protein.

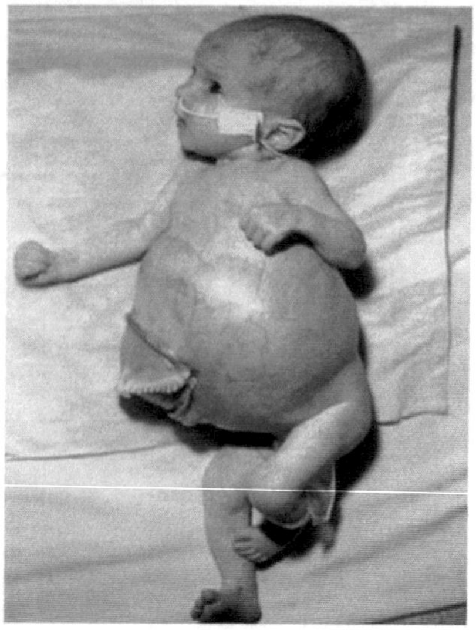

Bresler et al, Science Translational Medicine, 2011

Structure-Function Mapping of the Mutational Landscape

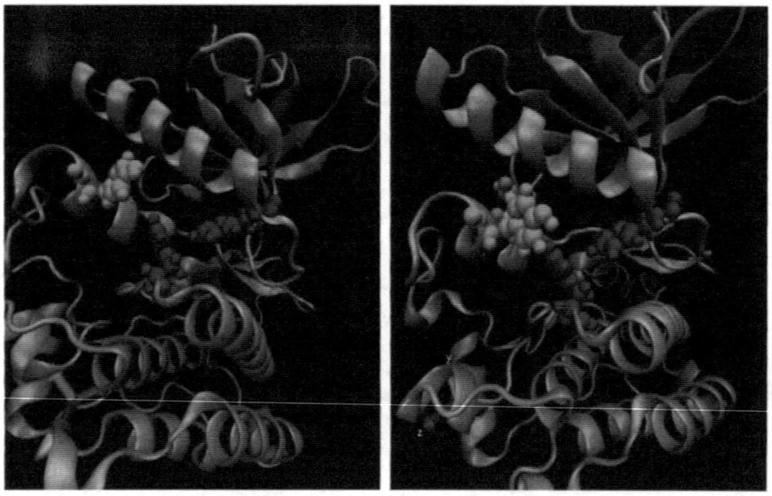

WILDTYPE R1275Q

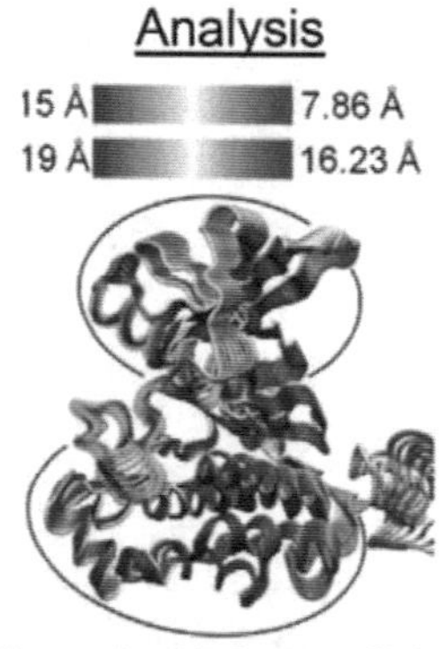

Hbonds (A-loop, C-helix)
Hydrophobic (SASA)
PCA

Shih, A.J., et al. Cancers, 2011; Biochem J, 2011; Telesco et al. Mol Biosyst, 2011

Insilico Oncology: Profiling of Kinase Mutations to Identify Drivers amongst Passenger Mutations

Multiple methods of mutation driven activation exist.

Mutations proximal to the A-loop and C-Helix appear to have very profound affects on activation.

The most common mutations (F1174L, F1245C, F1245V, R1275Q) all destabilize the hydrophobic core. The R1275Q mutation disrupts crucial hydrophilic interactions stabilizing the inactive structure.

Free energy studies will be required to fully understand destabilization effects. Much more can be learned by modeling ATP + Mg^{2+} or competitive inhibitors.

Computer simulations prove useful in predicting the effect of mutational landscape on kinase activation mechanisms (ErbB and Alk)

Shih, A.J., et al. Cancers, 2011; Biochem J, 2011; Telesco et al. Mol Biosyst, 2011; Weiser, Cancer Cell, at review

University of Pennsylvania Department of Bioengineering

Role of Modeling in Engineering and Biology

- Unlike fields like theoretical physics, modeling in biology and engineering do not stand alone
- Models have to be custom-built for specific applications/contexts and with support of experimental (in vitro, in vivo, biophysical) data
- Can such models be predictive? Yes, if they are properly validated across multiple scales
- Models can help identify core mechanisms through construction of minimal models
- Models can define crucial structural, kinetic, thermodynamic constraints
- Predictive Models can enable design

Thank You!

Group Members
- Ryan Bradley
- ~~Rick Tourdot~~
- E Joe Jordan
- Peter Huwe
- David Slochower
- Whelton Miller
- N. Ramakrishnan
- Hsiu-Yu Yu

- Andrew Shih
- Yingting Liu

Acknowledgments

Collaborators
- Mark Lemmon
- John Maris
- Yael Mosse
- Matthew Lazzara
- Kate Ferguson
- Portonovo Ayyaswamy
- David Eckmann
- Vladimir Muzykantov

Undergraduates
- Josh Weinstein
- Fei Jia
- Sean Engles
- Randall Toy
- Vyas Ramanan

Funding
- NSF CBET

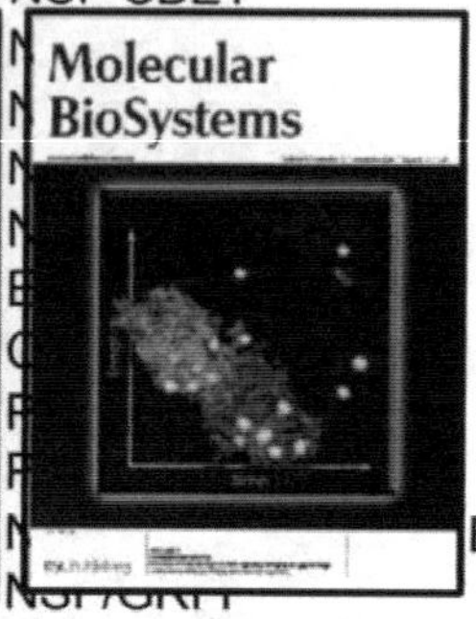

BMB

NSF/GRFP

- Ashton Fellowship

University of Pennsylvania

Department of Bioengineering

UNIVERSITY of ROCHESTER

Fingerprint-guided engineering of selective P450 catalysts for sp^3 C—H functionalization in complex molecules

Rudi Fasan

Department of Chemistry

University of Rochester

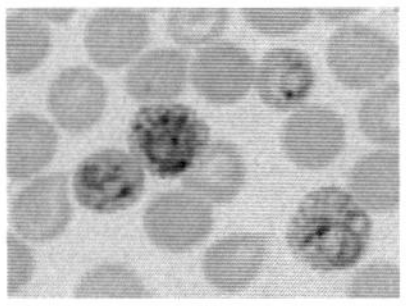

Fasan group

- *Synthesis and evolution of Macrocyclic Organo-Peptide Hybrids (MOrPHs) as modulators of protein-protein interactions*

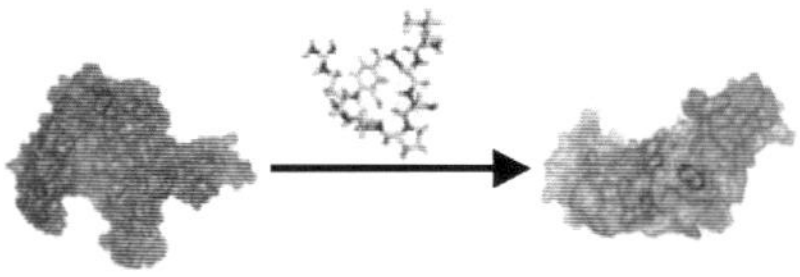

Smith et al, *Angew .Chemie Int. Ed.* (**2011**) 50(22):5075-80
Satyanarayana et al, *Chem. Comm.* (**2012**) 48(10):1461-3
Frost et al, *ChemBioChem* (**2013**), 14(1):147-160
Smith et al, *J. Org. Chem.* (Synopsis to appear in 2013)

- *P450-centered methodologies for late-stage sp^3 C—H functionalization in complex molecules*

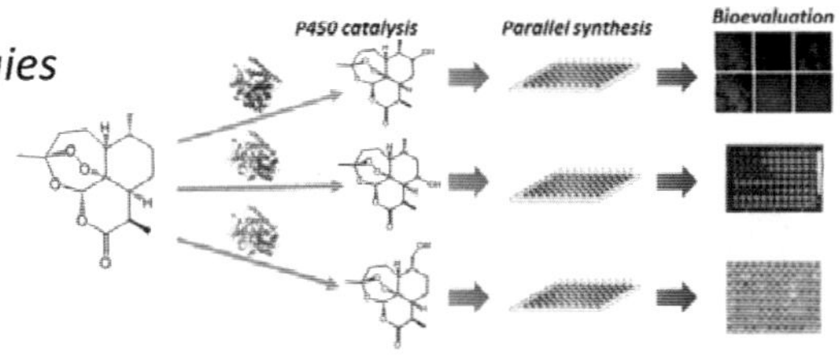

Aliphatic C-H bonds

Aliphatic (sp^3) C—H bonds are ubiquitous in bioactive natural products and synthetic molecules

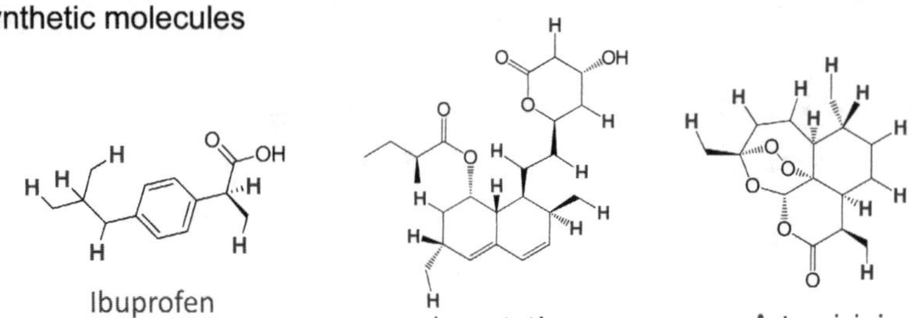

Ibuprofen
(antiinflammatory)

Lovastatin
(cholesterol lowering drug)

Artemisinin
(antimalarial drug)

➢ Methods for selective functionalization of sp^3 C—H bonds can offer broadest range of opportunities for elaboration of a complex organic scaffold

➢ **Key challenges :**
- *strong chemical bonds (BDE ≈ 100 kcal mol^{-1})*
- *small difference in reactivity*

Selective oxidation of aliphatic C—H bonds

➢ *Breslow, Ellman, Bergman, Sames, Sanford, Stahl, Suslick, Crabtree, Du Bois, Hartwig, Fuchs, Wender, Grieco, White, Baran, ….*

- ❑ transition metal catalysts + directing groups
- ❑ stoichiometric oxidizing reagents
- ❑ supramolecular enzyme mimics

Efficient catalytic methods that can provide tunable control on the oxygenation of unactivated C—H bonds are missing

- ➢ late-stage (FG compatibility)
- ➢ not requiring directing groups
- ➢ catalytic
- ➢ tunable selectivity (also towards 'unbiased' positions e.g. I° C–H bonds)
- ➢ efficient (high TON)

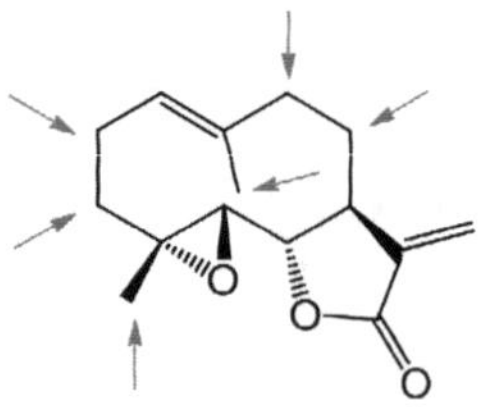

Cytochrome P450 monooxygenases

Catalyze the insertion of an oxygen atom into nonactivated C-H bonds under mildest reaction conditions (r.t., water) and using air oxygen

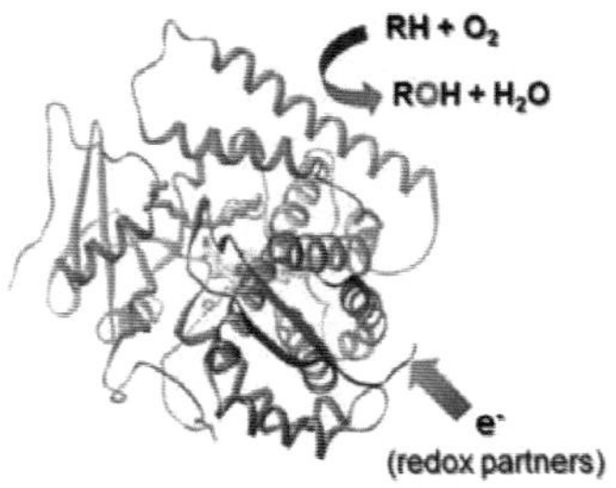

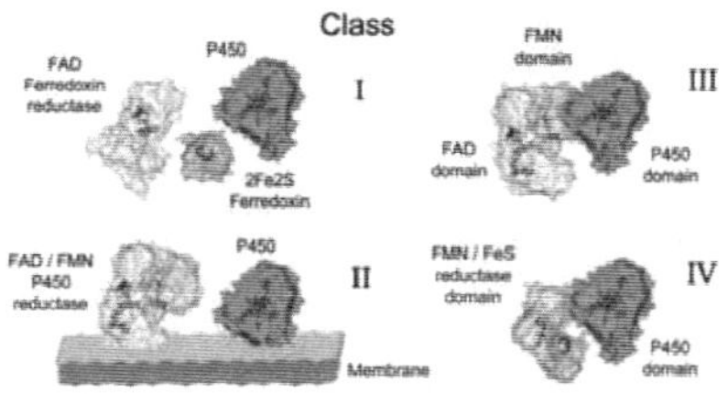

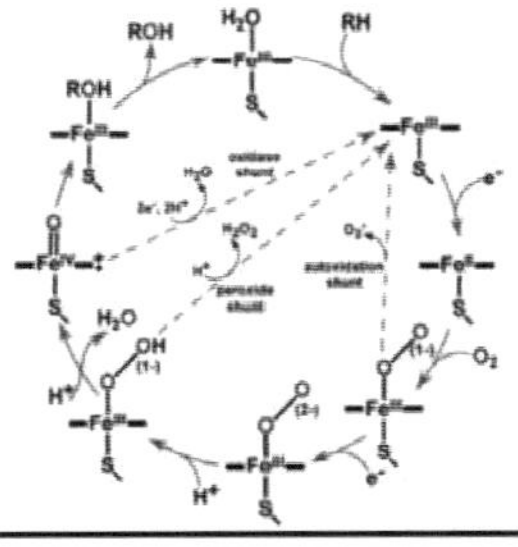

'Self-sufficient' P450s as versatile catalytic systems for chemical synthesis:

- Reductase fused to monooxygenase domain
- CYP102A1 ($P450_{BM3}$), CYP102A5, P450 RhF
- Soluble, high expression yield in *E. coli*
- Highly active (10-1000x eukaryotic P450s)

Exploring the 'C(sp^3)—H space' via P450-mediated synthesis

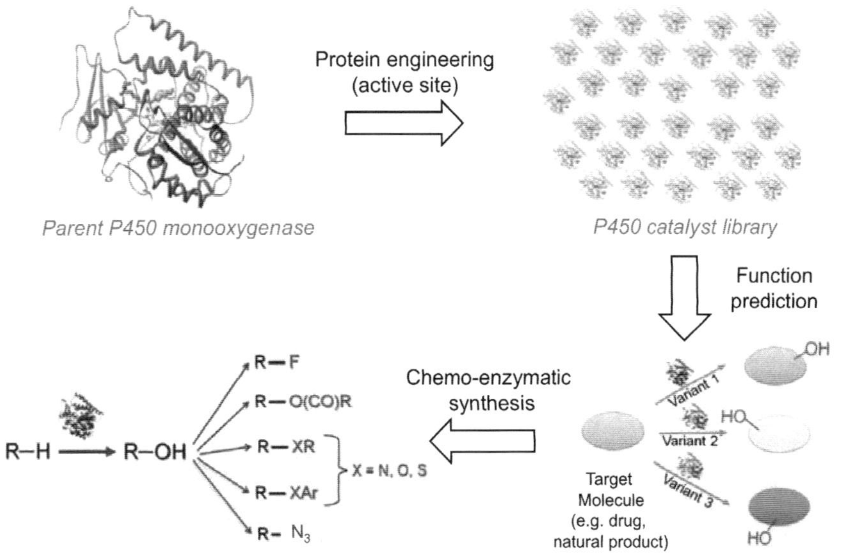

Chemo-enzymatic fluorination of small molecules

- ➢ H→F substitutions affect pharmacokinetic and pharmacological properties of drugs (bioavailability, *in vivo* half-life, metabolic stability, potency)
- ➢ 30% drugs on the market or in clinical trials contain fluorine

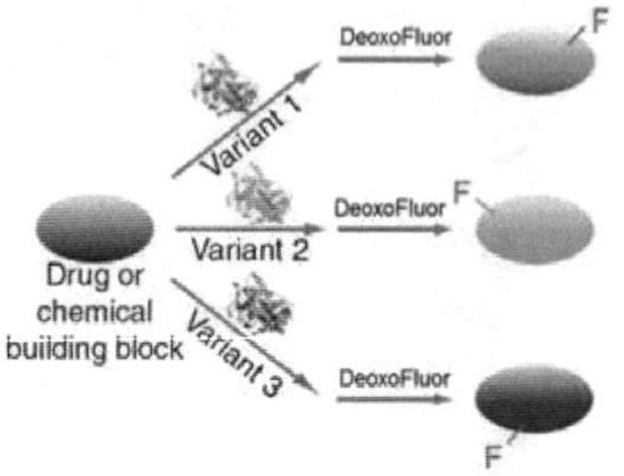

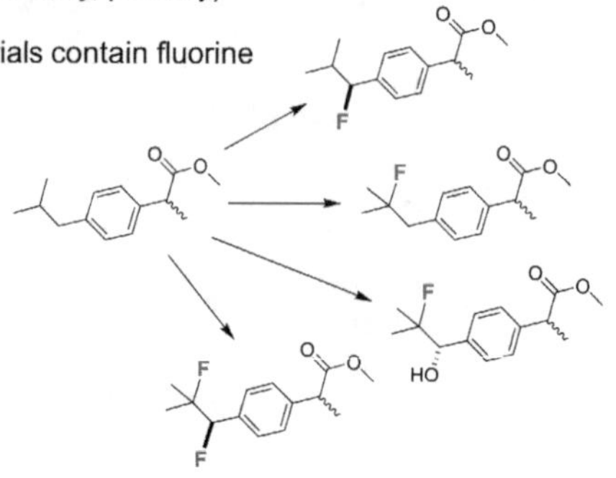

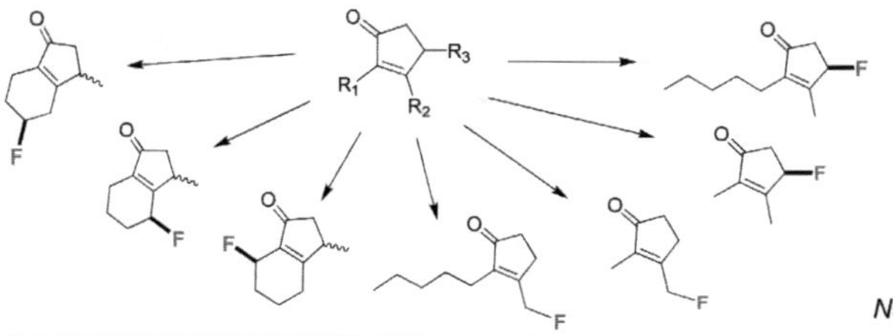

- 15 different substrates
- >0.3 g scale
- 35-90% selectivity
- 20-90% *ee*

Nature Chem. Biol. (**2009**) 1 (5): 26-28

Ibuprofen fluorination

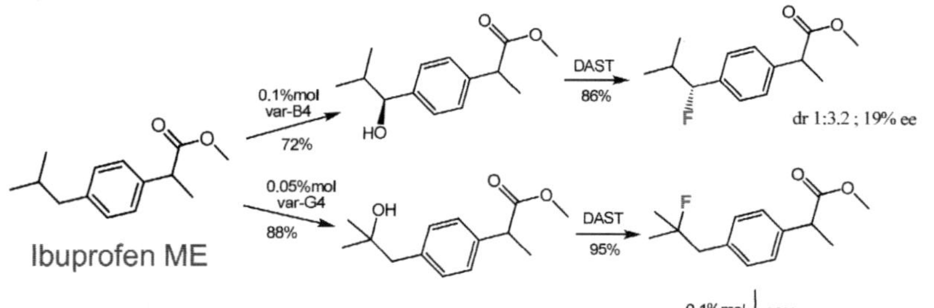

- 0.3 g scale
- 0.05-0.1 mol% P450
- >65% yield over 4 steps (difluorinated Ibu)

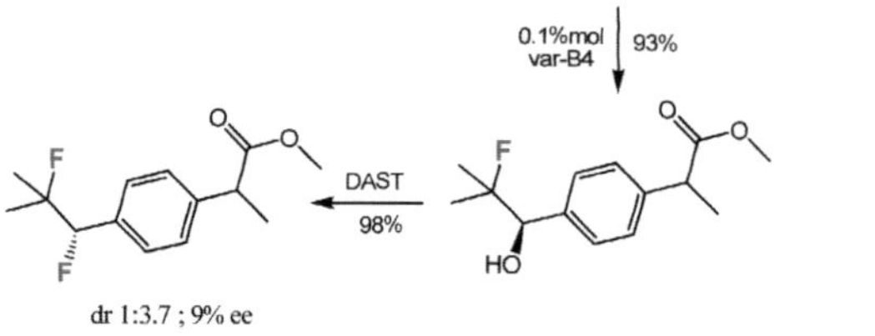

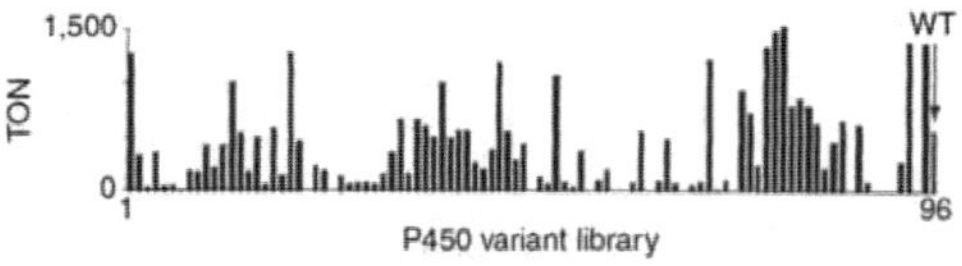

- ➢ 30% library show >500 TON
- ➢ 66% active subset: >70% regiosel.

Nature Chem. Biol (2009) 1 (5): 26-28

BBB permeability

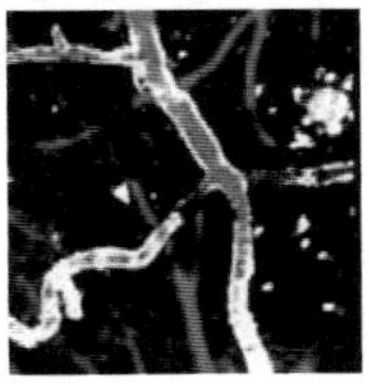

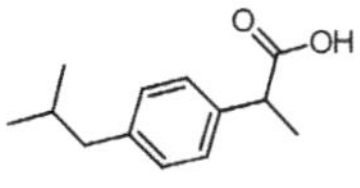

Ibuprofen lowers brain levels of highly amyloidogenic isoform amyloid β42 peptide (Weggen *et al.* (2001) *Nature* **414**, 212-216).

Aβ42-lowering compounds with improved brain permeability serve as prophylactic agents against Alzheimer's disease and related pathologies (Leuchtenberger *et al.* (2006) *Curr. Pharm. Des.* **12**, 4337-55)

➢ Parallel Artificial Membrane Permeability Assay (PAMPA) using blood-brain barrier mimic

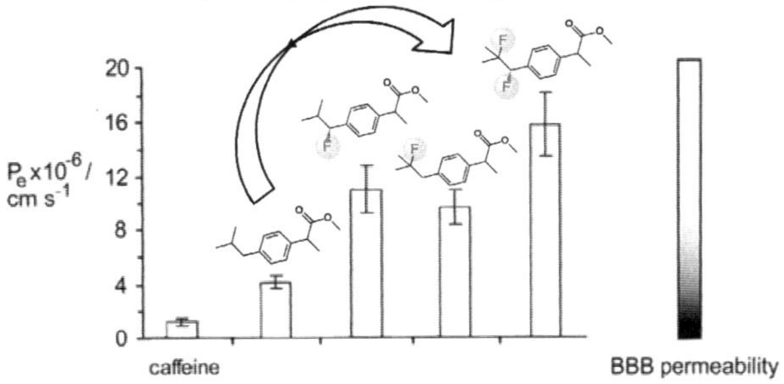

Nature Chem. Biol (2009) 1 (5): 26-28

P450 catalyst discovery process: challenges

Major bottleneck in exploiting P450 catalysis for synthesis is identification of P450 catalysts with desired substrate reactivity and regio/stereoselectivity

→ Compared to other properties, fine-tuning regio- and stereoselectivity in P450 enzymes has been remarkably challenging:

1) Structure-based rational 'design' had limited success due to plasticity of the active site and conformational changes upon catalysis

2) regio/stereoselectivity determination requires laborious, low-throughput analyses (GC/HPLC) *(e.g. 10^4 library on 5 substrates = 1 yr analytical time)*

3) Mutations that broaden substrate specificity often result in P450 catalysts with low site-selectivity

4) P450 variants with high regio/stereoselectivity are rare (1/10,000 or less)

Fasan, *ACS Catalysis* (2012) 2: 647-666

P450 active site configuration

Shape and geometry of the heme pocket in P450 enzymes dictates:

- ❑ active site accessibility ➢ Substrate reactivity
- ❑ orientation of substrate above heme prior to oxidative attack ➢ Regio/stereoselectivity

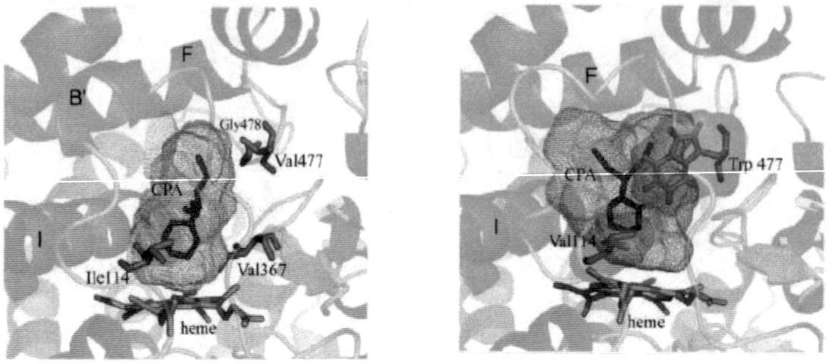

P450 functional fingerprinting

Concept: use set of chromogenic probes based on diverse scaffolds to map P450 active site shape and geometry and predict its reactivity via analysis of its "fingerprint"

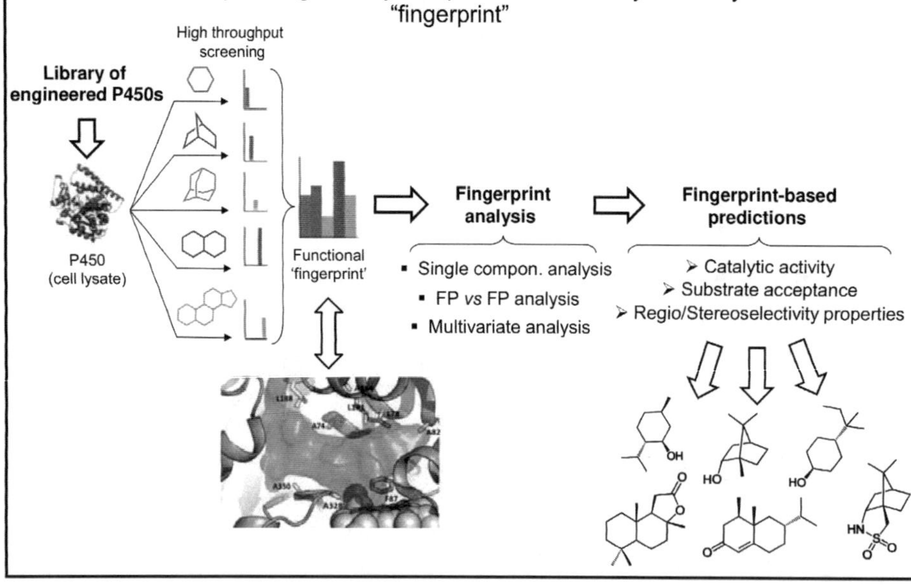

Fingerprint probes

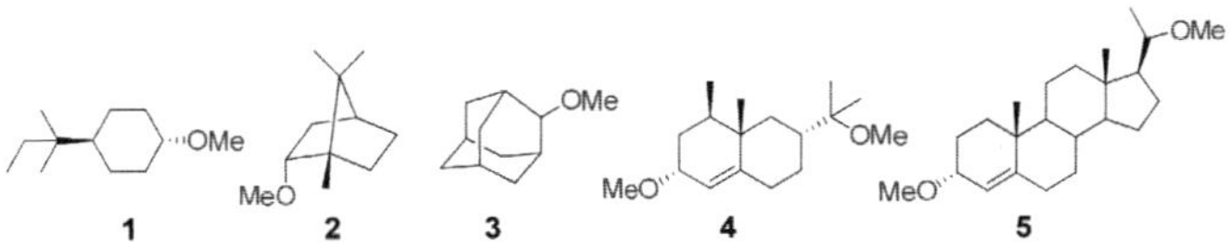

1 2 3 4 5

- *diverse 3D structures*
- *conformationally rigid*
- *gram scale synthesis from inexpensive s.m.*

➢ *'reporter' functional group* (—OCH_3):

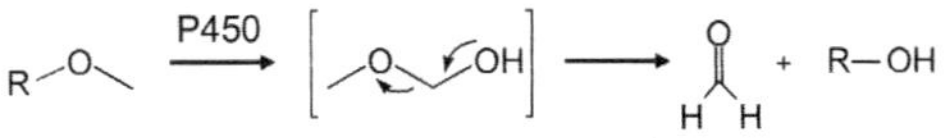

(detected colorimetrically)

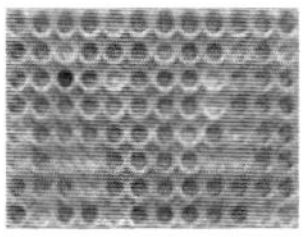

- *HT colorimetric assay*
- *cell lysate*
- *couples screen read-out with C—H oxidation*

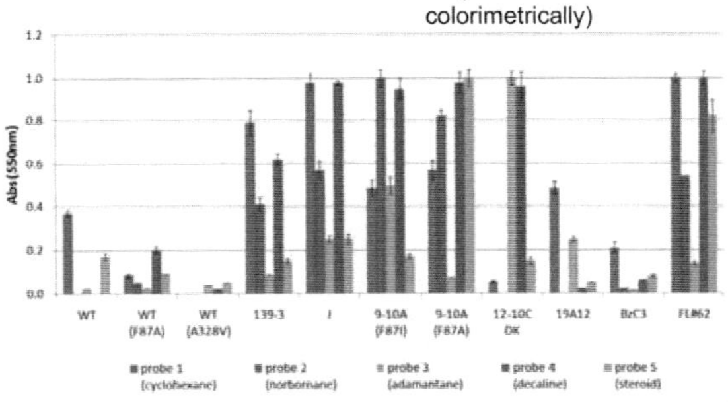

❑ each variant is associated to a unique fingerprint

❑ WT *vs* 139-3 (30 Å^3-larger heme cavity)

❑ probes **1**, **2**, **3** are within 20 Da but varying activity across P450 set

J. Am. Chem. Soc. (**2011**) 133(10):3242-5

P450 function prediction: substrate reactivity

Substrate acceptance is dictated by compatibility of substrate molecular scaffold with enzyme active geometry

➢ **hypothesis**: the individual FP components can be used as predictors of P450 reactivity toward probe-related substrates ("single FP component analysis")

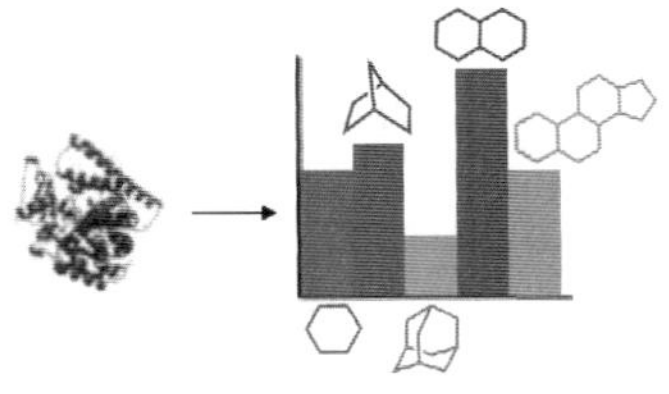

Target substrate:

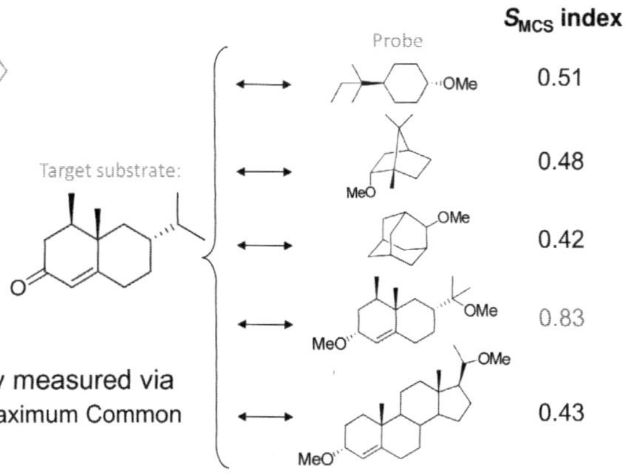

Target substrate-probe similarity measured via chemical similarity algorithm (Maximum Common Substructure index)

P450 function prediction: substrate reactivity

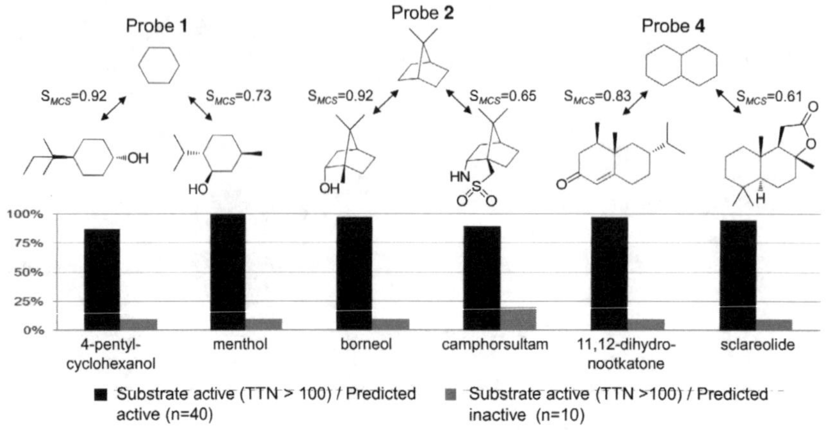

On average, 26% >1,000 total turnovers, 60% >700, and 78% > 400

85-100% accurate *a priori* predictions of P450 reactivity toward probe-related substrates via single fingerprint component analysis

J. Am. Chem. Soc. **(2011)** 133(10):3242-5

P450 function prediction: regioselectivity properties

Regio/stereoselectivity in P450-catalyzed oxidations is dictated by active site configuration

- **Hypothesis**: whole fingerprint profile can relay information about active site geometry and thus about regio/stereoselectivity properties of the fingerprinted P450s

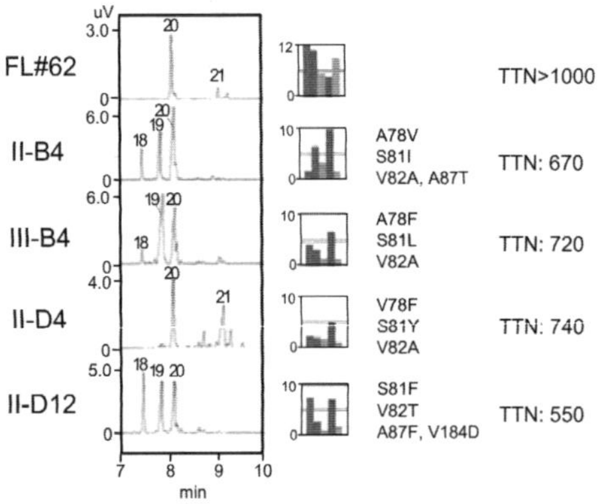

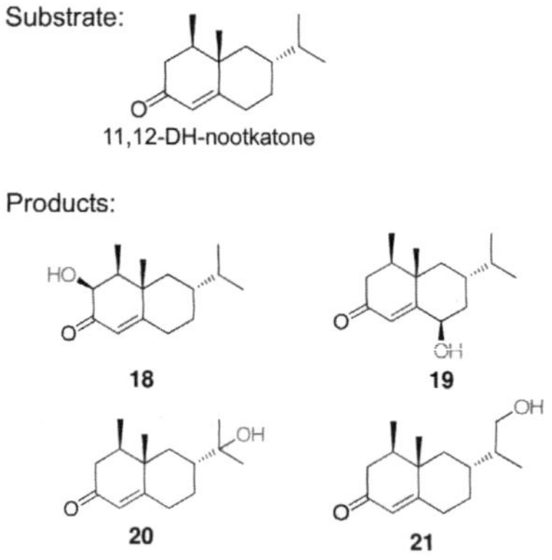

J. Am. Chem. Soc. **(2011)** 133(10):3242-5

P450 function prediction: regioselectivity properties

- ❑ across 6 substrates, ~40% of variants with divergent FP show different regio/stereoselectivity
- ❑ out of 40 P450s tested, 40% of the sp^3 C—H sites in the substrates could be targeted, including III°, II°, and even I° positions
- ❑ targeted aliphatic positions located across the entire carbon skeleton of the substrate

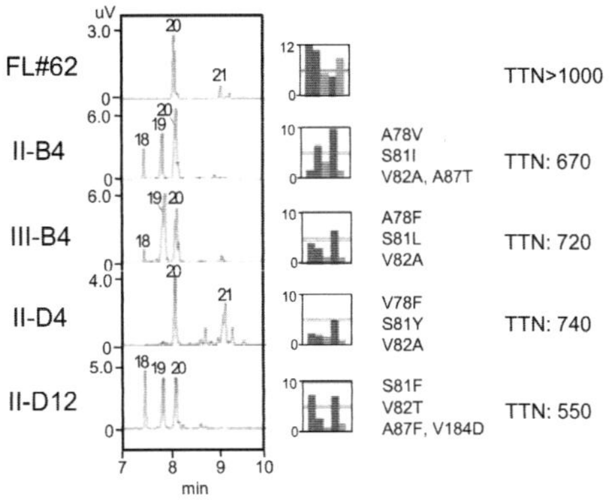

Substrate:

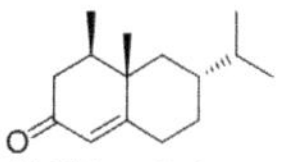

11,12-DH-nootkatone

Products:

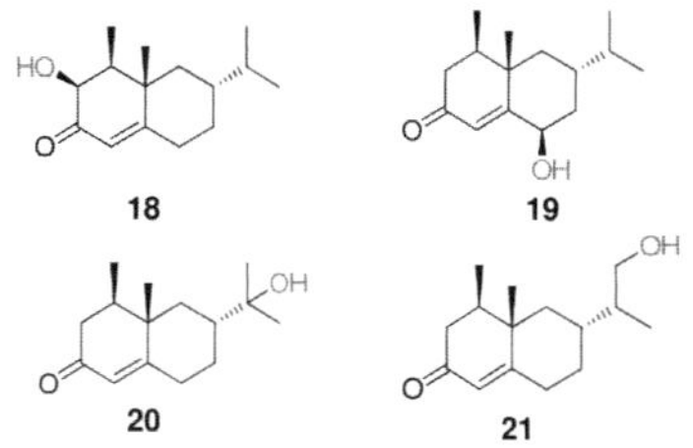

J. Am. Chem. Soc. (**2011**) 133(10):3242-5

P450 function prediction: regioselectivity properties

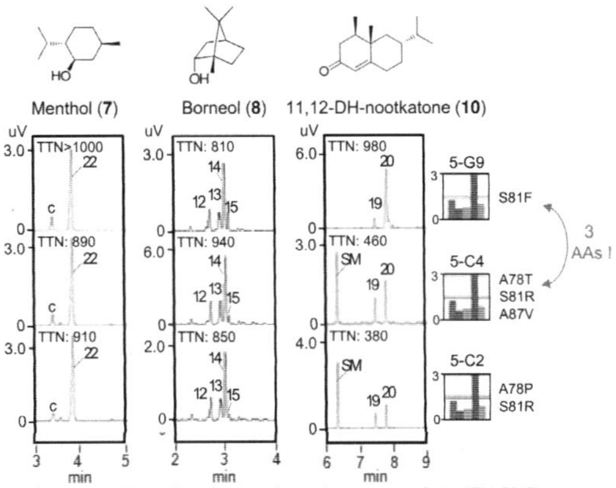

Subs.	P450	Products									
		22	c*								
7	5-G9	89%	11%								
7	5-C4	91%	9%								
7	5-C2	93%	7%								
				12	13	14	15	16			
8	5-G9			12%	10%	69%	9%				
8	5-C4			12%	10%	76%	2%				
8	5-C2			12%	10%	76%	2%				
									18	19	20
10	5-G9									15%	85%
10	5-C4									30%	70%
10	5-C2									31%	69%

- ➢ FP analysis can aid:
  - ➢ identification of 'functional replicants' in a P450 library (neutral mutations)
  - ➢ search for P450 variants with a specific type of regio/stereoselectivity once corresponding FP profile is known

J. Am. Chem. Soc. (**2011**) 133(10):3242-5

Scope

- The core structures of the probe series are shared by hundreds of naturally occurring terpenes

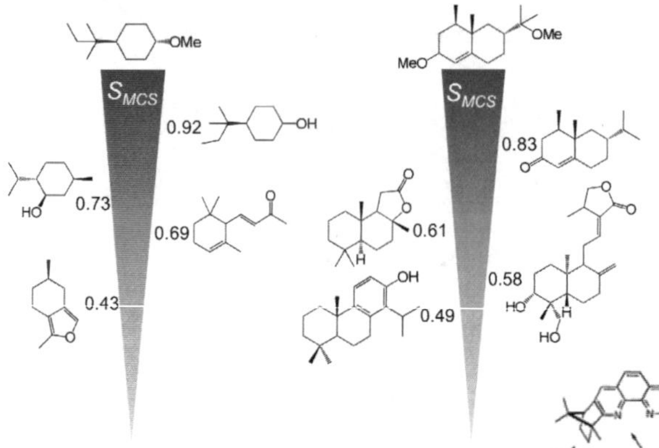

Decaline scaffold (probe **4**) constitutes the backbone of members of the labdane, cadinane, drimane, eudesmane, and aristolane subclasses, many of which have relevant biological activity

- elaboration of terpene scaffolds for generation of chiral catalysts for asymmetric transformations

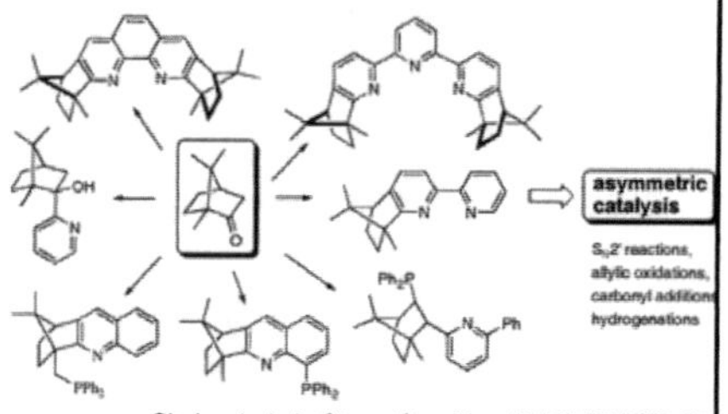

Chelucci *et al.*, *Chem. Soc. Rev.* (2006) **35**:1230-43

Next challenges

1) Can we expand the scope of FP-based P450 reactivity predictions to probe-unrelated substrates?

2) Can we use this strategy to streamline the development of P450 catalysts with fine-tuned regio- and stereoselectivity?

A. annua

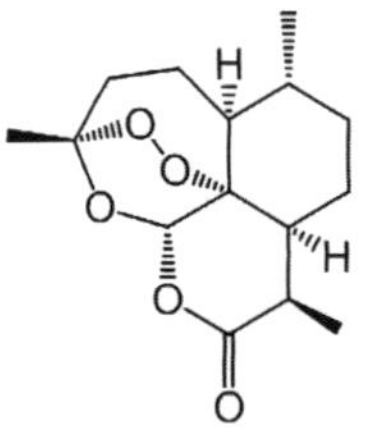

Artemisinin
(antimalarial drug)

Artemisinin (ART)

Malaria problem: 500 Mio infections per year causing > 1.5 Mio deaths

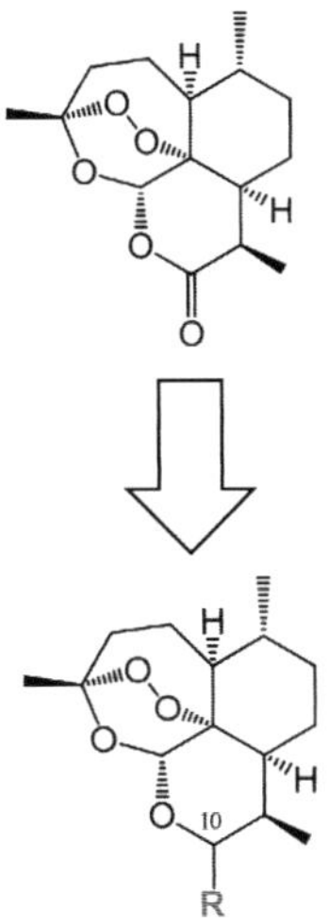

- fastest acting antimalarial drug
- life-saving solution against MDR *P. falciparum* strains (ACTs)
- low toxicity, high therapeutic index

- Very short half-life (< 1 hr) → high dosage → high-cost/limited supply
- 2007: first *Pf* isolates with reduced sensitivity in Cambodia

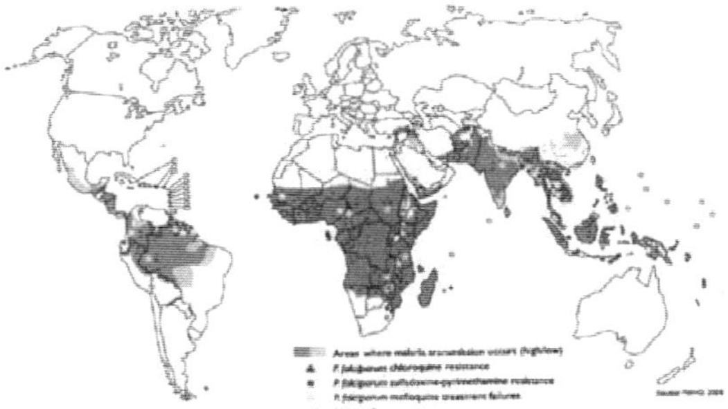

White, *Science* (2008) 320: 330-4

ART metabolism

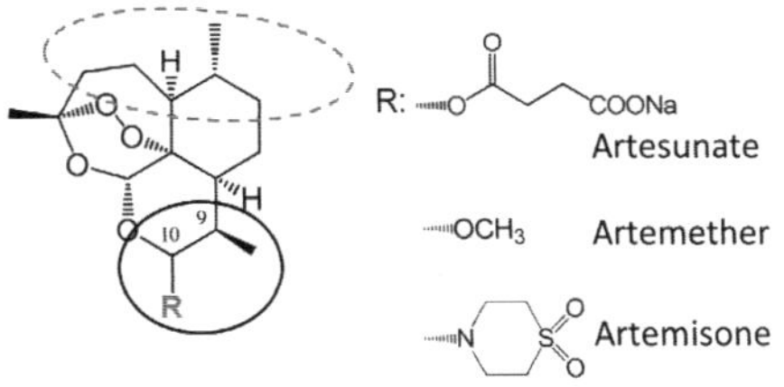

Chemical methods to functionalize ART are confined to C9 and C10

Chaturvedi *et al., Chem. Soc. Rev.* (2010) 39, 435-454

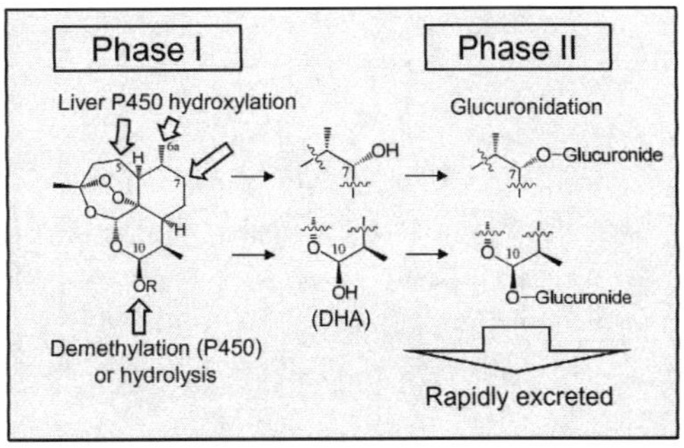

Hypothesis: Modification of upper hemisphere (C7, C6a, C5) could prevent or slow down metabolism in humans allowing for lower dosage, reduced ACT costs, increase availability of ACTs,…

O'Neill *et al., J. Med. Chem.* (2004) 47, 2946-60

Approach: P450-mediated C-H functionalization

➢ Starting point: P450 variant with promising ART-hydroxylating activity but only poor selectivity

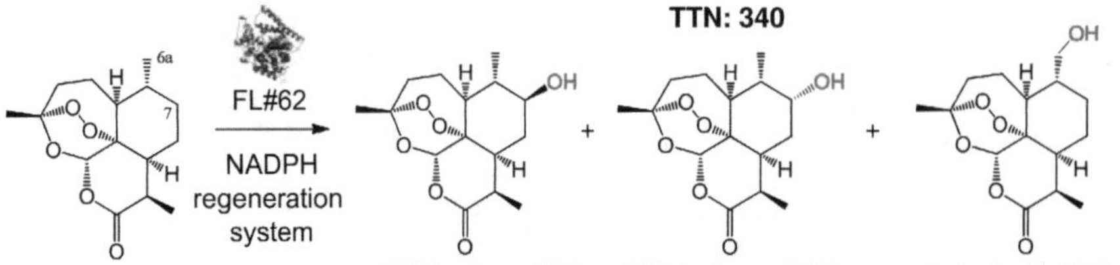

7(*S*)-hydroxy-ART	7(*R*)-hydroxy-ART	6a-hydroxy-ART
87%	8%	5%

J. Am. Chem. Soc. (**2012**) 134(45):18695-704

Tier 1: Create P450 catalyst diversity via 'first-sphere' active site mutagenesis

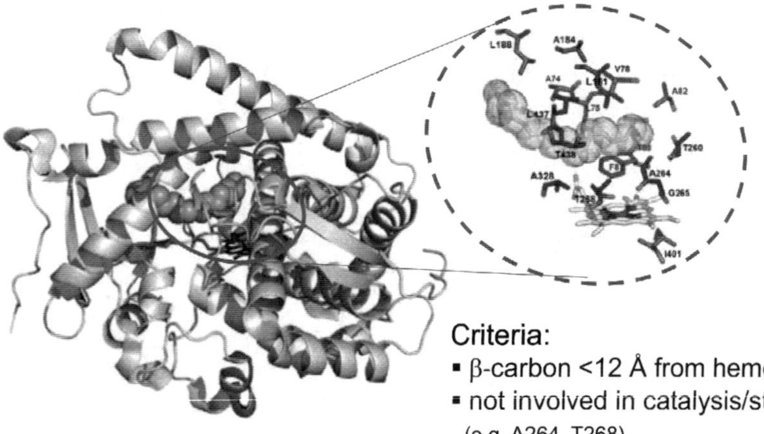

CYP102A1 variant FL#62

Criteria:

- β-carbon <12 Å from heme iron
- not involved in catalysis/structural role (e.g. A264, T268)
- simultaneous site-saturation (NNK) mutagenesis at multiple sites (2, 3, 4)

Tier 2: Identify functionally diverse P450 catalysts via fingerprint comparative analysis

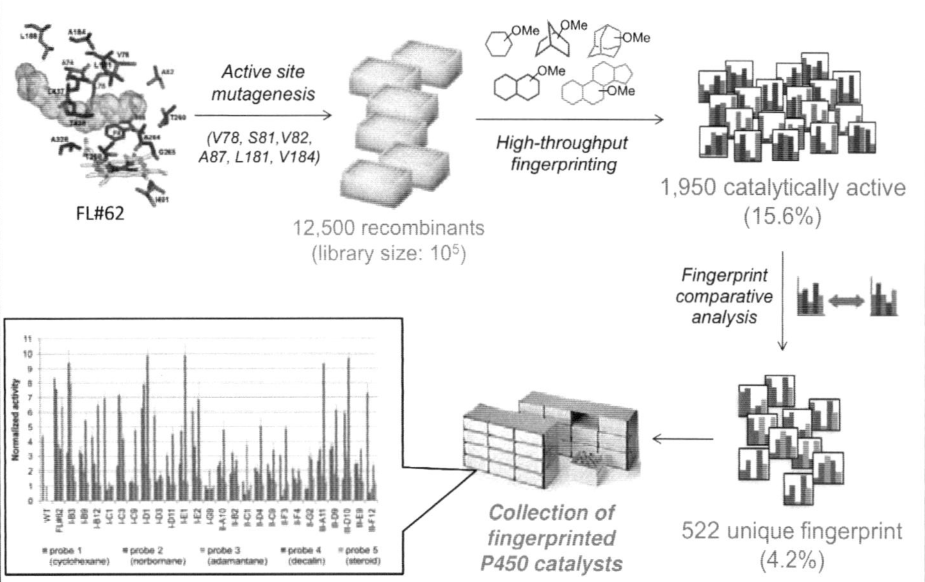

J. Am. Chem. Soc. (**2012**) 134(45):18695-704

Tier 3: ART reactivity prediction via FP multivariate analysis

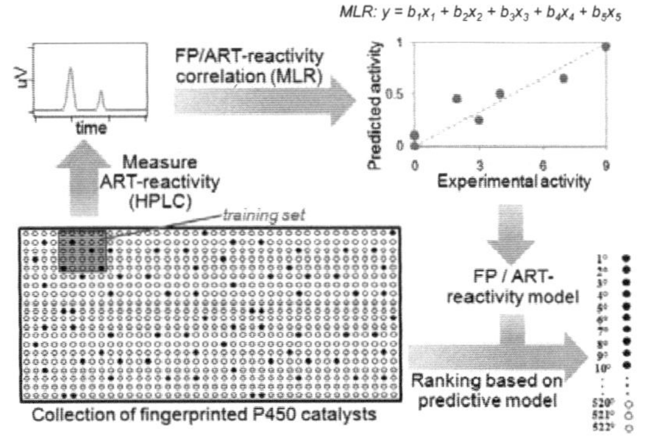

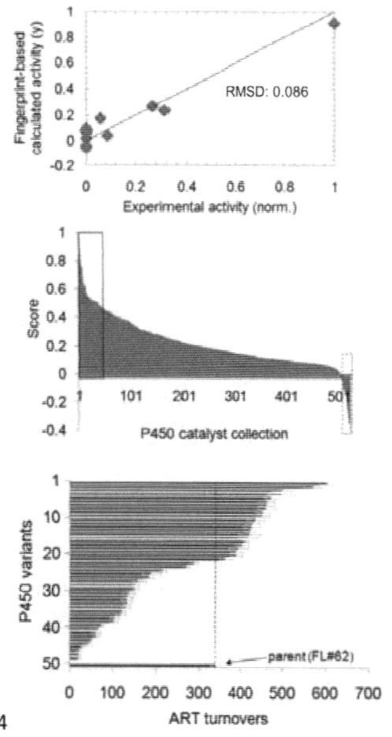

- ❑ 78% predicted ART-active (39/50) show high activity (>100 TTN; AVG: 323 TTN), half of which show higher TTN than FL#62
- ❑ 90% predicted ART-inactive (18/20) show no activity

J. Am. Chem. Soc. (**2012**) 134(45):18695-704

Regio- and stereoselective P450s for C7 oxidation

Variant	Product distribution			% de (C7)	TTN
	7(S)-OH-ART (2)	7(R)-OH-ART (3)	6a-OH-ART (4)		
FL#62	*83%*	*10%*	*7%*	*78% (S)*	*339*
IV-H4	100%	0%	0%	100% (S)	362
I-A1	100%	0%	0%	100% (S)	49
I-E2	100%	0%	0%	100% (S)	21
V-H2	96%	0%	4%	100% (S)	434
III-H2	95%	0%	5%	100% (S)	61
IV-C11	93%	0%	7%	100% (S)	454
I-G12	92%	0%	8%	100% (S)	413
I-H4	92%	0%	8%	100% (S)	498
I-D12	92%	0%	8%	100% (S)	287
VI-C2	91%	0%	9%	100% (S)	29
I-B12	91%	0%	9%	100% (S)	497
I-C12	89%	0%	11%	100% (S)	136
III-A11	89%	11%	0%	79% (S)	569
I-B4	89%	0%	11%	100% (S)	184
II-H10	0%	100%	0%	100% (R)	270
III-B1	19%	81%	0%	62% (R)	403
I-B1	20%	75%	5%	55% (R)	388
II-E1	49%	51%	0%		426
III-B7	51%	49%	0%		75
V-D8	53%	47%	0%		428
IV-E10	53%	47%	0%		152
IV-C3	58%	42%	0%		422
II-H2	56%	40%	4%		67
II-E2	22%	30%	48%		393
II-A2	38%	29%	34%		21
III-F1	46%	21%	33%		45

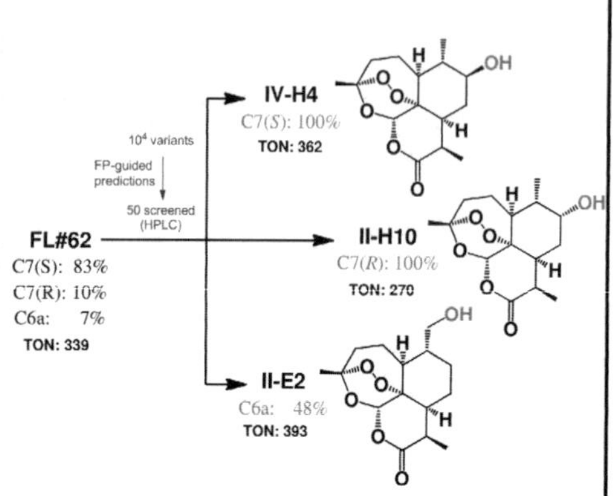

J. Am. Chem. Soc. (**2012**) 134(45):18695-704

Refining C6a regioselectivity

II-E2 found to carry 3 active site mutations (A78N, S81F, V82A) leaving additional first-sphere residues available for mutagenesis

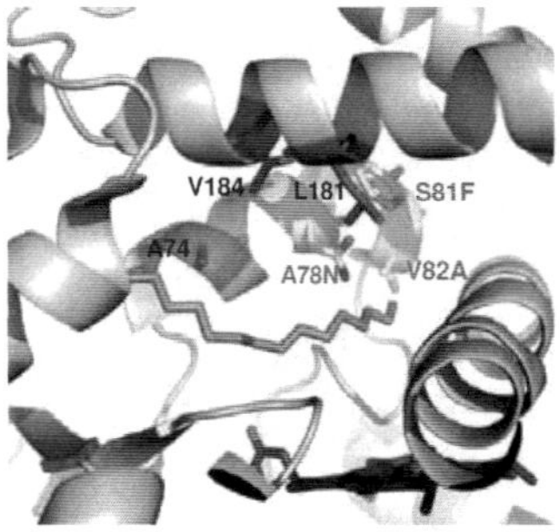

P450 II-E2

↓ *first-sphere AS mutagenesis*

74/181/184 NNK library

↓ *HT fingerprinting (3,000)*

120 functional variants (4%)

↓ *FP vs FP analysis*

50 unique-fingerprint P450s (1.6%)

↓ *FP-based prediction of ART reactivity*

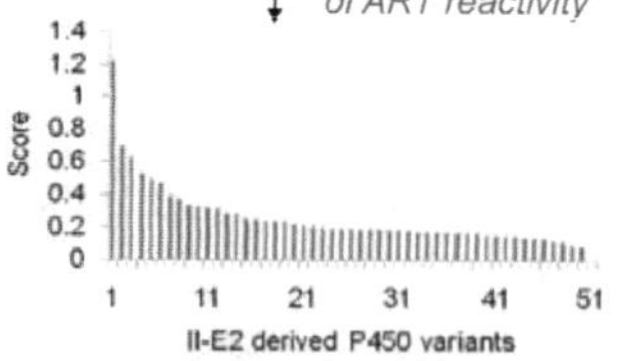

Refining C6a regioselectivity

Variant	Product distribution			TTN	StdDev
	7(*S*)-OH-ART (2)	7(*R*)-OH-ART (3)	6a-OH-ART (4)		
II-E2	*22%*	*30%*	*48%*	*393*	*25*
[II-E2] II-G1 (="X-E12")	4%	2%	94%	113	12
[II-E2] III-E3 (="X-F11")	0%	8%	92%	376	19
[II-E2] III-D2	0%	8%	92%	46	2
[II-E2] XIV-B9	5%	5%	90%	87	6
[II-E2] VI-B10	0%	12%	88%	149	12
[II-E2] VI-H4	3%	12%	86%	70	4
[II-E2] II-E12	6%	9%	84%	115	17
[II-E2] XVII-E10	11%	8%	81%	171	9
[II-E2] II-D10	0%	23%	77%	105	11
[II-E2] VI-F4	11%	27%	63%	45	6
[II-E2] XII-F10	0%	43%	57%	88	8
[II-E2] II-A5	16%	33%	51%	58	7
[II-E2] XIII-B2	91%	0%	9%	474	23
[II-E2] VII-D3	0%	16%	84%	86	2
[II-E2] XV-E12	6%	13%	80%	59	9
[II-E2] VI-A2	0%	21%	79%	126	8

Top half of library: rows [II-E2] II-G1 through [II-E2] XIII-B2. Bottom half of library: rows [II-E2] VII-D3 through [II-E2] VI-A2.

- ❑ 81% of ART-active are within 25 top-scoring variants
- ❑ 90% of predicted ART-inactive variants (9/10) show no activity
- ❑ Most 6a-selective variants (92-94% regioselect.) within 10 top-ranking P450s

J. Am. Chem. Soc. (**2012**) 134(45):18695-704

Regio- and stereoselective P450 catalysts for ART hydroxylation

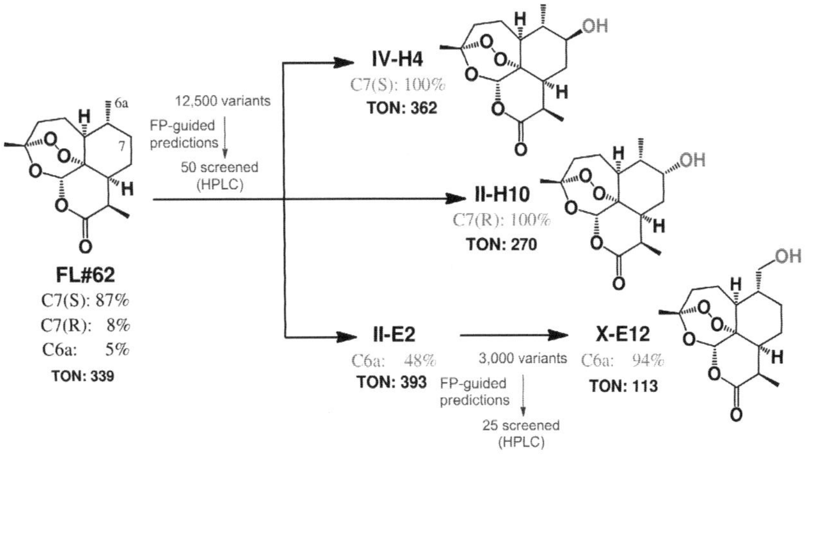

J. Am. Chem. Soc. (**2012**) 134(45):18695-704

Functional characterization

Variant	Amino acid substitutions							Product (%)			TON	K_D (μM)	Product formation rate	Coupling efficiency (%)
	74	78	81	82	87	181	184	7(*S*)-OH	7(*R*)-OH	6a-OH				
FL#62	A	A	S	V	A	L	V	83	10	7	339 ± 12	29 ± 5	316 ± 20	41.9
IV-H4		S	V	A				**100**	0	0	362 ± 15	53 ± 11	100 ± 4	71.4
V-H2		S	I	A				**96**	0	4	434 ± 21	45 ± 10	41 ± 2	23.5
II-H10		N	F	T	F	F	T	0	**100**	0	270 ± 8	61 ± 2	32 ± 2	12.8
III-B1		F	F	A				19	**81**	0	403 ± 17	38 ± 2	72 ± 2	23.3
X-E12	V	N	F	A		A	T	4	2	**94**	113 ± 12	300 ± 24	72 ± 3	45.1
X-F11	T	N	F	A			S	0	8	**92**	376 ± 19	234 ± 29	56 ± 2	56.3

❑ Mutational distance reflects functional distance from parent enzyme. Functionally related variants converged to similar mutational patterns.

❑ ART binding affinity within range of wt CYP102A1 for native substrates (K_D: 10-300 μM)

❑ General trade-off between site-selectivity and product formation rate

❑ TTN and % coupling comparable or higher than unselective parent enzyme

Selective P450 catalysts for ART oxidation

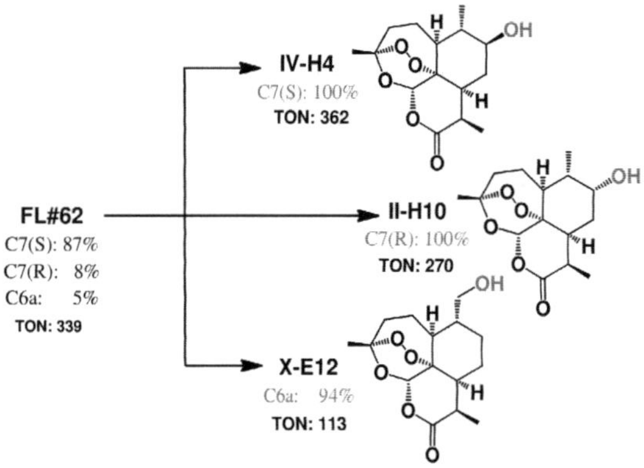

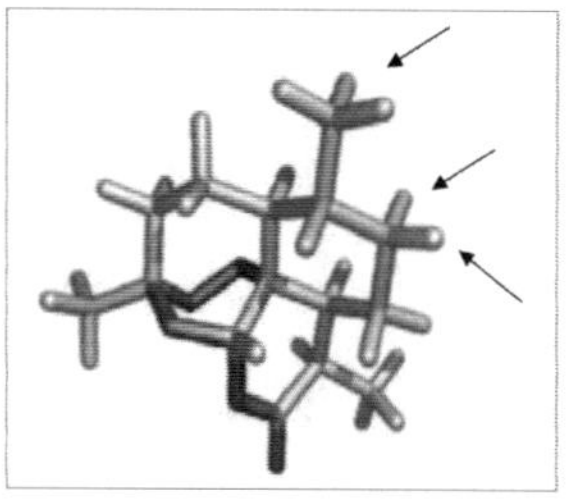

Acknowledgements

Postdocs
Dr. Natalie Bordeaux
Dr. Ritesh Singh
Dr. Francesca Vitali
Dr. Kaidong Zhang*

Graduate students
Abby Cryan
John Frost
Joshua Kolev
Jessica Smith

Undergraduates
Philip Cistrone
Tatyana Dyndikova
Nick Jacob
Yick Chong Lam
Louis Papa
Shady El Damaty*
Brian Shafer*
Matthew Demars*

* Former students

NIGMS
(R01 GM098628)

Chemistry of Life Processes
(CHE-1112342)

Provost's Multidisciplinary Research Award

Graduate/Undergraduate Fellowship Programs:
NSF-GRFP, McNair Program, NSF-REU, De Kiewiet Summer Program

Altering the Apoptotic Network in Cancer: A Promising Therapeutic Approach

Kurt Deshayes

Department of Early Discovery Biochemistry
Genentech Inc.

January 15, 2013

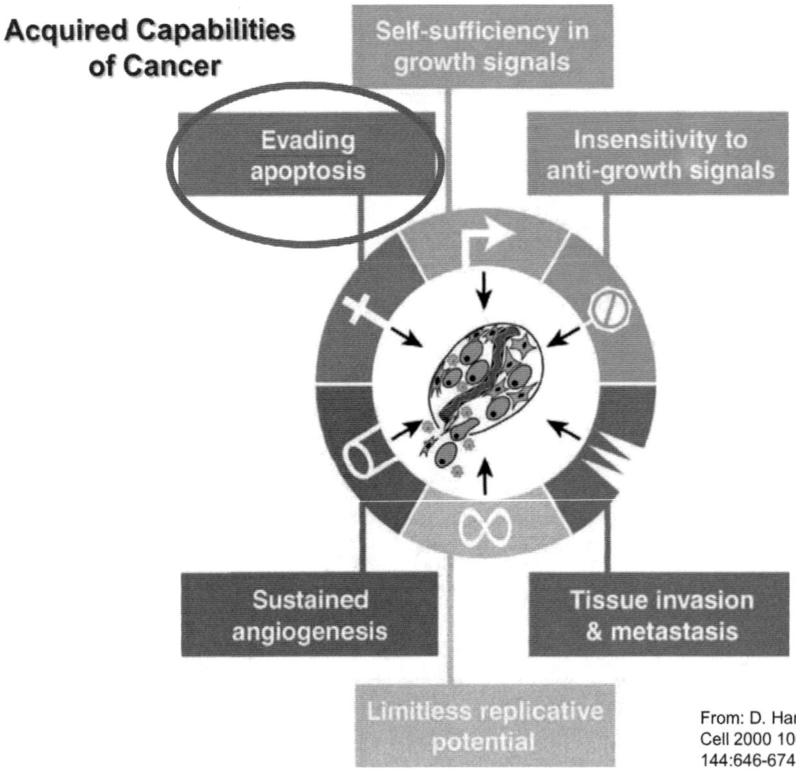

From: D. Hanahan, R.A. Weinberg, Cell 2000 100:57-70 and Cell 2011 144:646-674.

Apoptosis is the process by which damaged cells are removed.

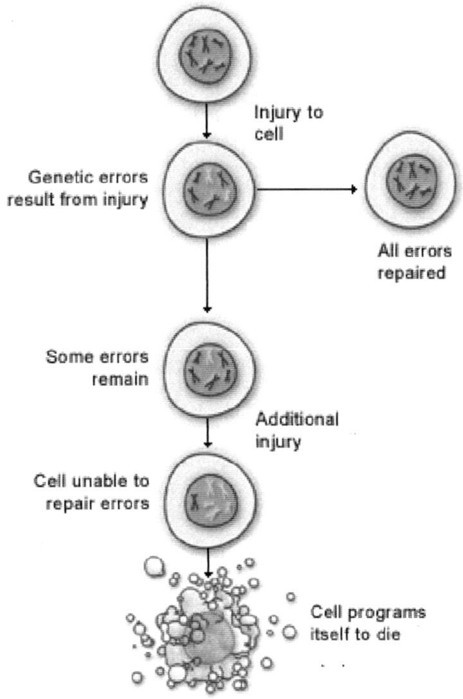

Apoptotic bodies are absorbed by phagocytes.

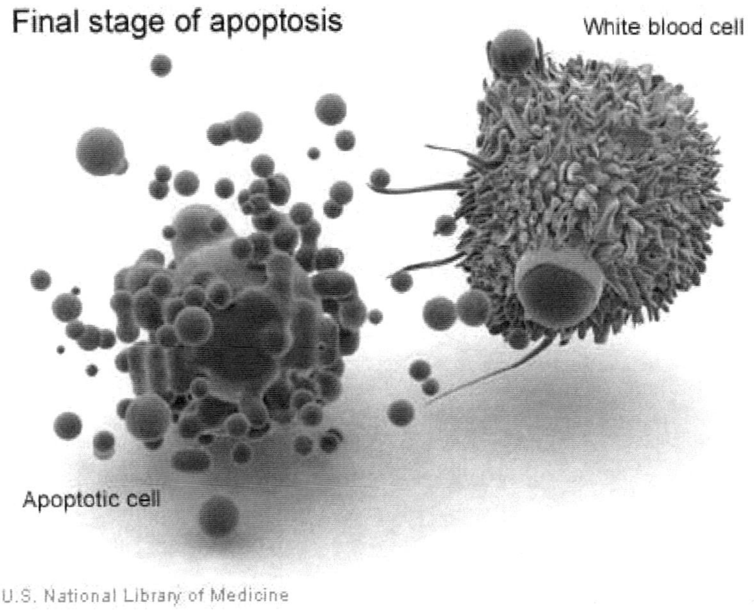

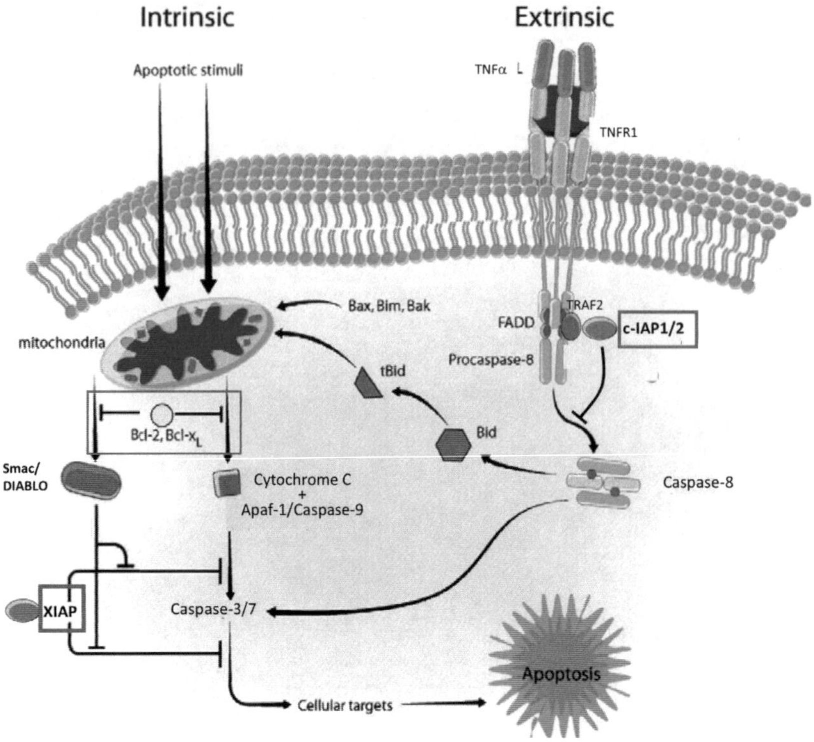

Central question: Can the Apoptotic pathway in Cancer be rewired using small molecules that modulate key protein-protein interactions.

1. Yes, but you need to pick the right target.

2. There is no general method that works in all cases.

IAP proteins are overexpressed in cancer cells

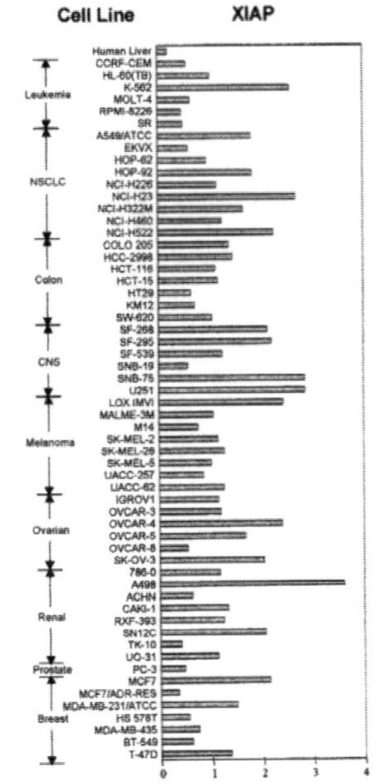

Fong et al. (2000) Genomics 70, 113

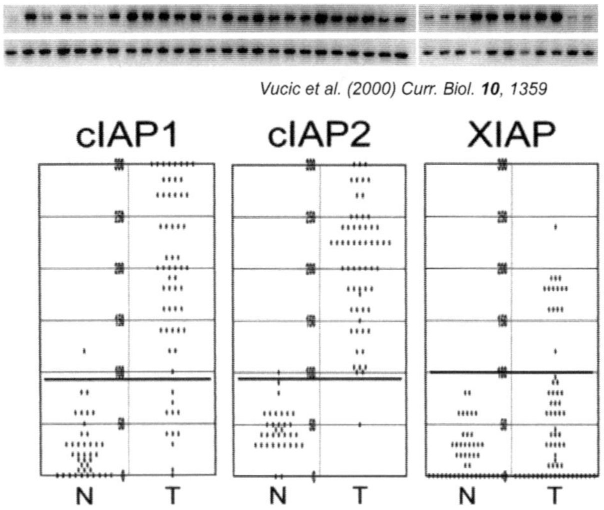

Vucic et al. (2000) Curr. Biol. **10**, 1359

Krajewska et al. (2003) Clan. Cancer Res. **9**, 4914

The IAP Protein: Compact Binding Site

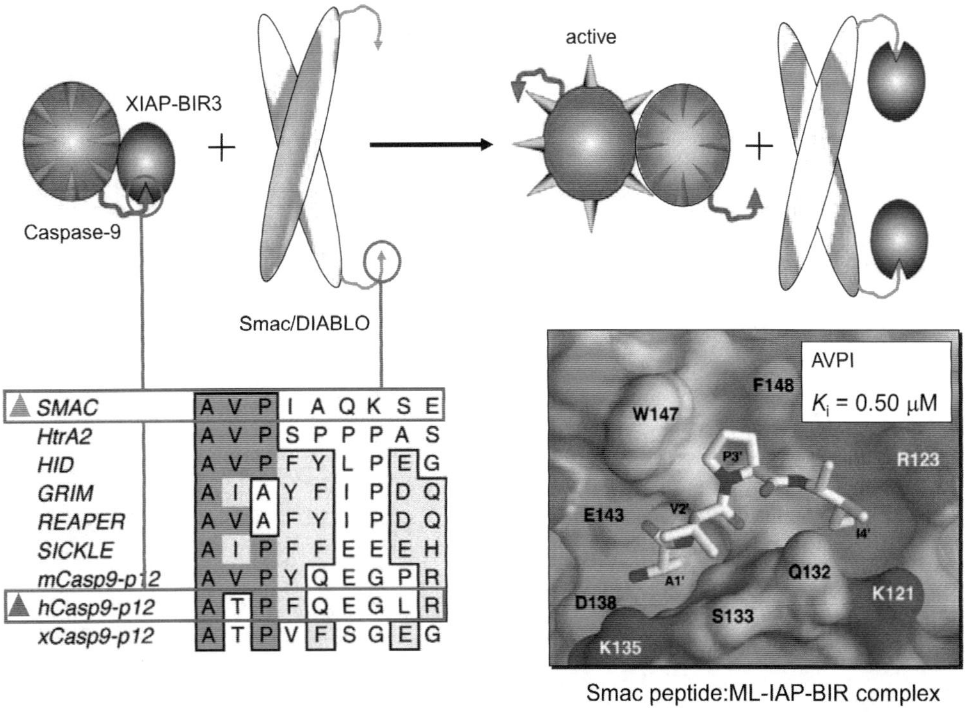

Smac peptide:ML-IAP-BIR complex

Peptide ligand was converted into Peptidomimetic GDC-0152

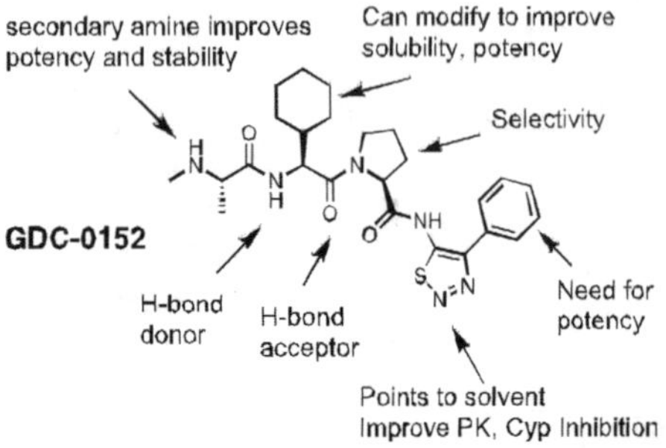

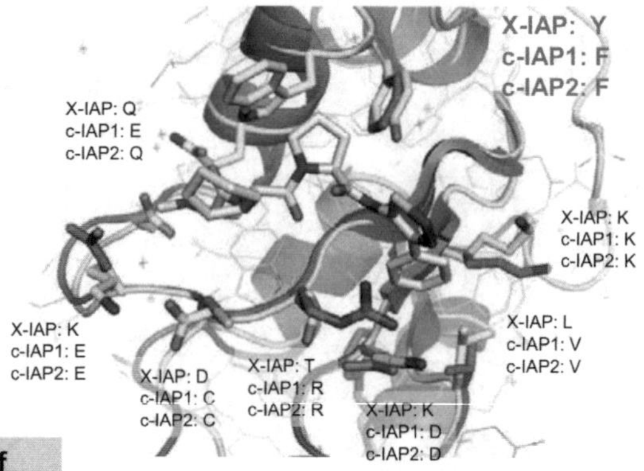

Structure driven synthesis and evaluation of hundreds of diverse compounds has determined the key structural drivers for stability, solubility, selectivity, Cyp inhibition, oral bioavailability, potency, and efficacy.

GDC-0152 bound to XIAP overlaid with c-IAP1

J. Flygare et al. J. Med. Chem., 2012, 55 (9), pp. 4101–4113

GDC-0152

- GDC-0152: a small molecule antagonist of inhibitor of apoptosis (IAP) proteins
  - block XIAP:caspase-9 interaction
  - block ML-IAP:Smac interaction
  - block c-IAP:Smac interactions
  - Potent cell-based activity demonstrated with several cancer cell lines but not with non-cancer cells (single agent or in combination with death ligands or standard chemotherapeutic agents)
  - Efficacious at inhibiting tumor growth in mouse xenograft models of breast cancer, NSCLC, pancreatic cancer, melanoma, and colon cancer

AVPIAQKSE

K_i(ML-IAP-BIR) = 0.50 μM
K_i(XIAP-BIR3) = 0.67 μM
K_i(XIAP-BIR2) = 13.8 μM
K_i(cIAP1-BIR3) = 0.08 μM
K_i(cIAP2-BIR3) = 0.17 μM

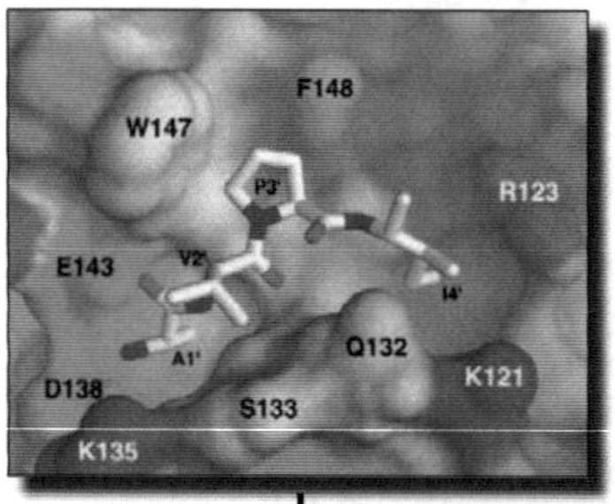

Structure-based evolution of peptidomimetic

GDC-0152
MW ~ 500 Da

K_i(ML-IAP-BIR) ≤ 0.009 μM
K_i(XIAP-BIR3) ≤ 0.021 μM
K_i(XIAP-BIR2) = 0.112 μM
K_i(cIAP1-BIR3) ≤ 0.017 μM
K_i(cIAP2-BIR3) = 0.043 μM

GDC-0152 prevent endogenous association of ML-IAP and Smac in melanoma cells.

SK-MEL28 melanoma cells

– –

GDC-0152 – –

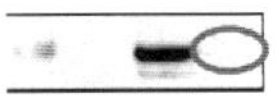

IP: αSMAC
W: αML-IAP

W: αML-IAP

W: αSMAC

D (10 μM) for 20 h

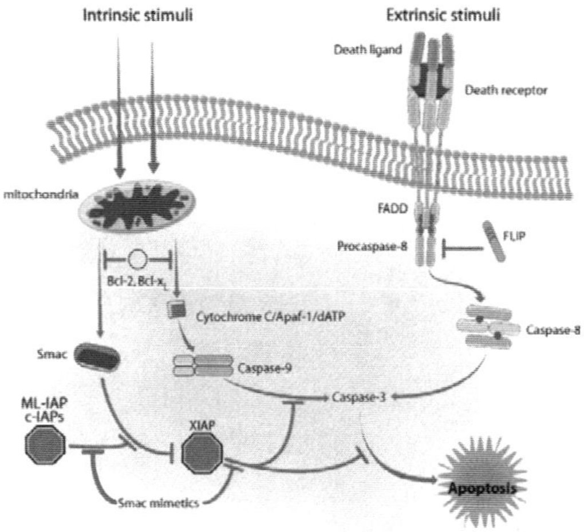

Domagoj Vucic

Single agent activity of GDC-0152 in human NSCLC xenograft tumors.

Subcutaneous A549 NSCLC xenograft tumors

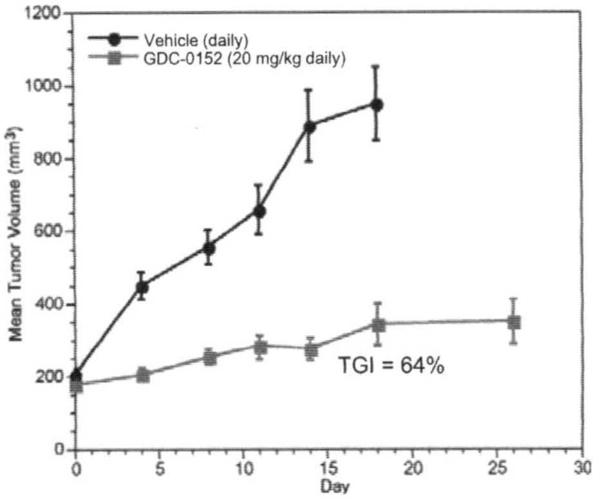

Orthotopic H460 NSCLC xenograft tumors

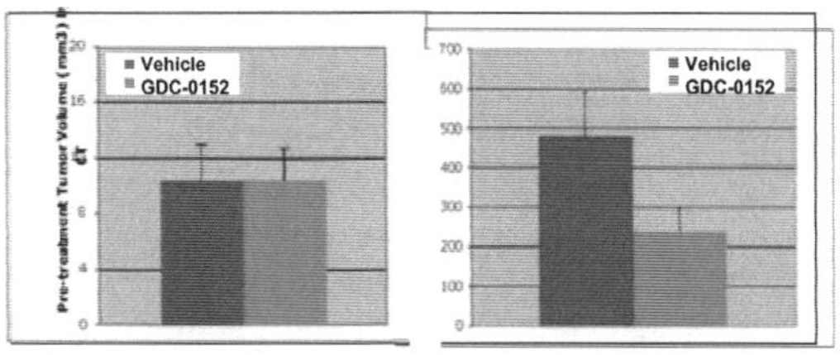

Micro-CT imaging

Pre-Treatment **Post-Treatment (2 wks)**

Vehicle

GDC-0152

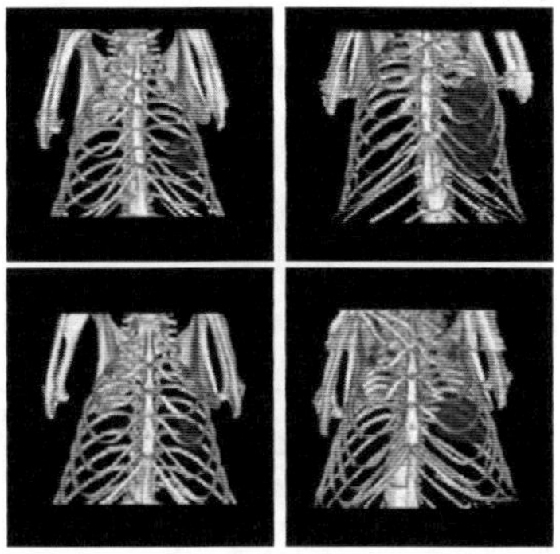

Karl Doerner, Lesley Murray, Hongkui Jin

GDC-0152 led to the Discovery of GDC-0917 which is currently undergoing human clinical trials.

1. A potent (< 1 μM) peptide ligand could be converted into more potent peptidomimetics with good drug like properties.

2. Once potent and selective molecules were in hand it becomes possible to investigate the pathway in details. For example the unique regulation of the ubiquitin ligase activity of cIAP. E. Varfolomeev **Cell** 2007 13: 669-81 and E. C. Dueber **Science** 2011 *334:376-380.*

The BCL-2 Family of Proteins

=> regulators of apoptosis

BCL-2 faction:	motifs:	function:	members:
Pro-Survival	BH4 BH3 BH1 BH2	block activation of apoptosis	Bcl-2 Bcl-x_L Mcl-1 Bcl-w
executioner	BH3 BH1 BH2	permeabilize outer MTC membrane	Bax Bak
damage sensors	BH3	neutralize anti-apoptotic member	Bad Bid Bim Noxa Puma

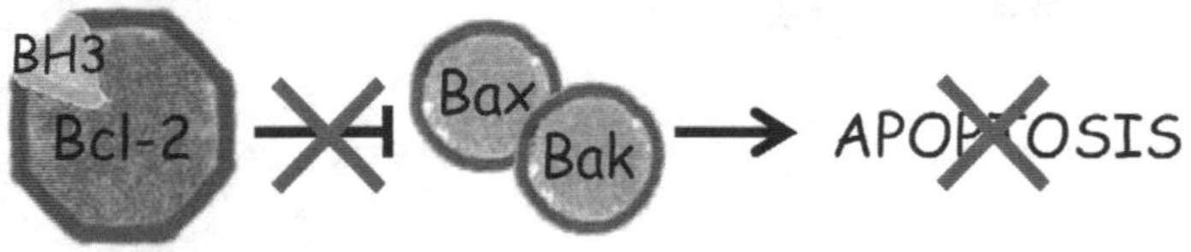

Bcl-2 Family Members as Drug Targets

1. Pro-survival members, particularly Bcl-2, Bcl-x_L, and Mcl-1 are overexpressed in cancers and facilitate chemotherapeutic resistance.

1. If only one could make small molecule BH3 mimetics:
   a. neutralize pro-survival proteins
   b. promote apoptosis

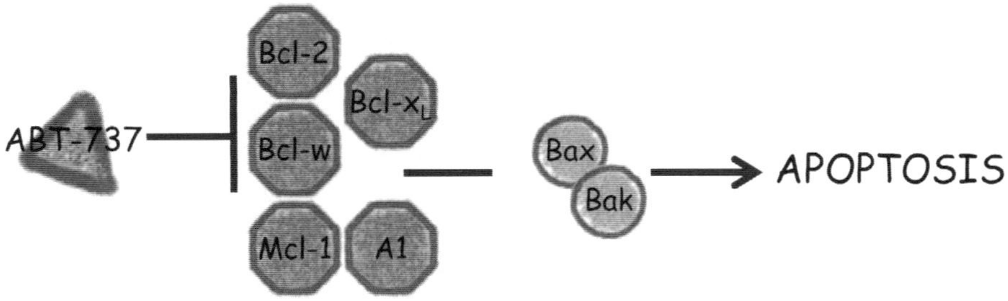

ABT-737 inhibits a broad spectrum of pro-survival family members, but not Mcl-1.

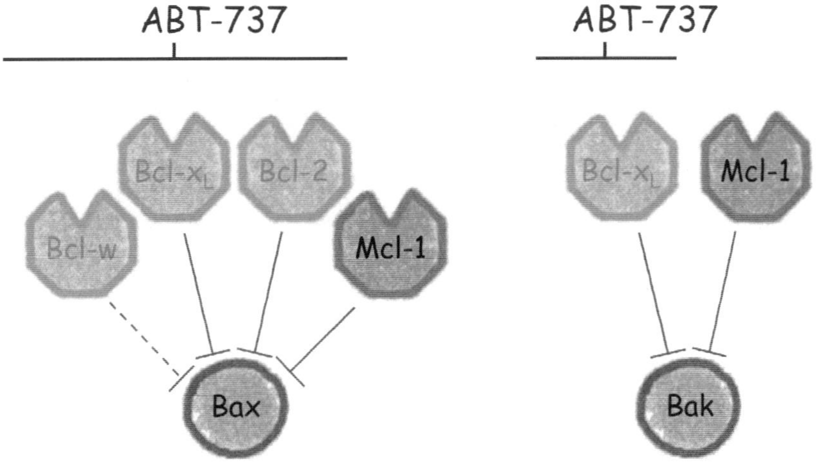

In contrast, Bim inhibits all the pro-survival family members.

T. Oltersdorf et al. Nature (2005) 435, 677-81

BH3 Mimetic: ABT-737 binds in the BIM groove within Bcl-x_L , not a compact binding site. Lead discovered using fragment screening.

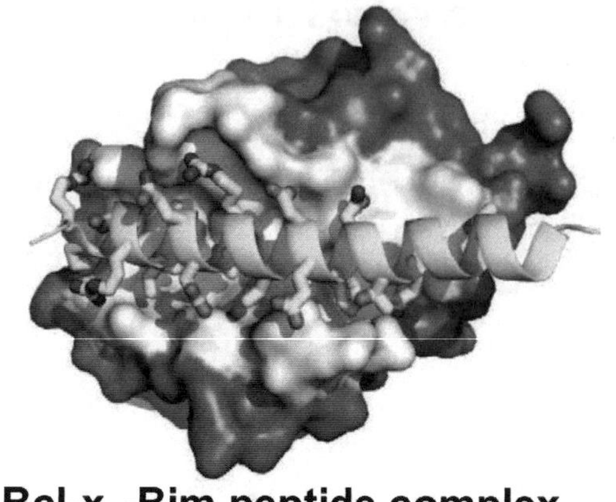

Bcl-x_L-Bim peptide complex

PDB 1PQ1
Liu et al., Immunity 2003

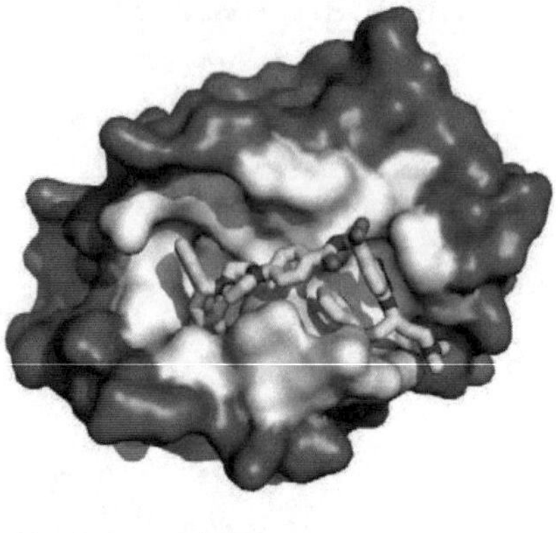

Bcl-x_L-ABT-737 complex

PDB 2YXJ
Czabotar et al., Cell Death Diff. 2007

Bcl-x_L Selective Program

Hypothesis: Correlation of solid tumor resistance with Bcl-x_L expression levels with122 standard chemotherapy agents in 60 NCI cell lines suggests solid tumors may be sensitized to combination therapy with a BCl-x_L antagonist.

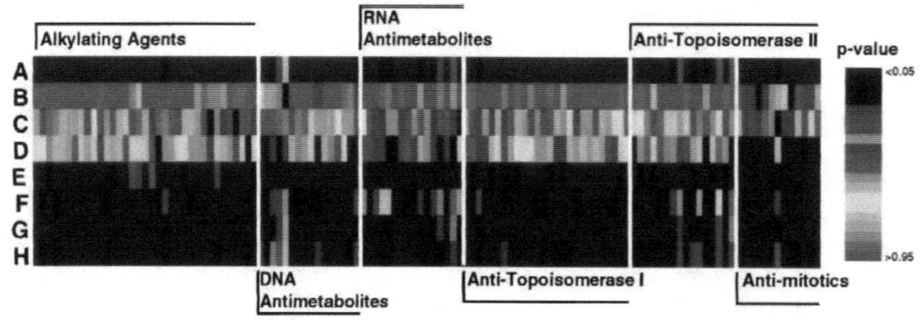

Amundson, Cancer Res.60, 6101-10 (2000)

A: Bcl-x_L
B: p53
C: Bcl-2
D: Bax
E: Doubling time
F: Bcl-x_L - doubling time
G: Bcl-x_L - p53
H: Bcl-x_L- p53 mutant lines

Question: Will a Bcl-x_L selective antagonist in combination with standard chemotherapy have equivalent efficacy with reduced tox compared to ABT-263 ?

Initial Screen at the WEHI in Melbourne

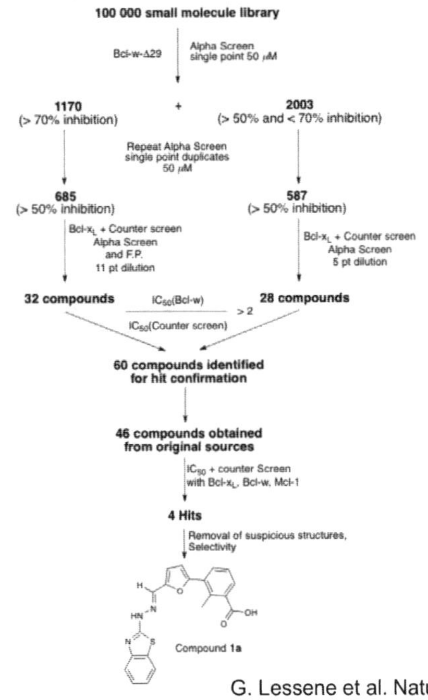

G. Lessene et al. Nature Chem Bio. Under Revision

Hit A was chosen due to its superior selectivity

Binding affinities (IC_{50} values) in μM

Hits	Bcl-w		Bcl-xl			Mcl-1		Counter screen
	Alpha screen	FP	Alpha screen	FP (Bim)	FP (Bak)	Alpha screen	FP	
A	33	>240	7.7	20	6.8	110	167	249
B	32	37	100	93	50	32	15	172
C	149	167	233	215	188	85	44	>300
D	38	23	30	22	32	5	5.4	24

Three compounds were rejected: weak or non-selective

Potency was improved 3000 fold to give compound 3

Original hit **A**: IC_{50} (Bcl-X_L) = 7.7 μM

↓ *12 compounds*

Compound **1**: IC_{50} (Bcl-X_L) = 2.2 μM

↓ *180 compounds*

Compound **2**: IC_{50} (Bcl-X_L) = 0.12 μM

↓ *130 compounds*

Compound **3**: IC_{50} (Bcl-X_L) = 0.023 μM

The Pyridine in 3 can be replaced by a Thiodiazole to give equal potent 4.

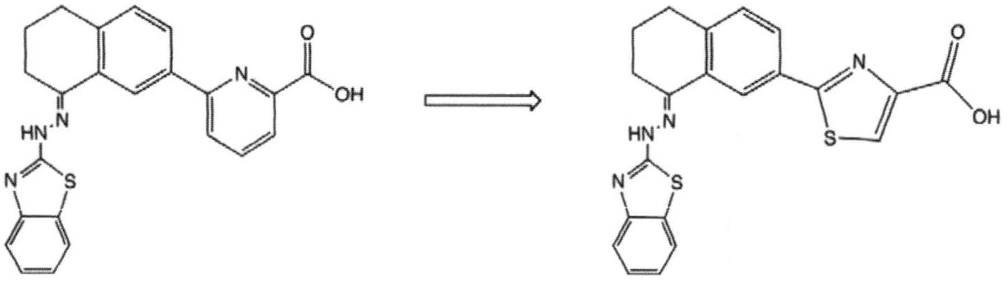

3 IC_{50} (Bcl-X_L) = 0.023 μM

4 IC_{50} (Bcl-X_L) = 0.024 μM

Extending compound 4 across the BH3 binding site Improves Potency.

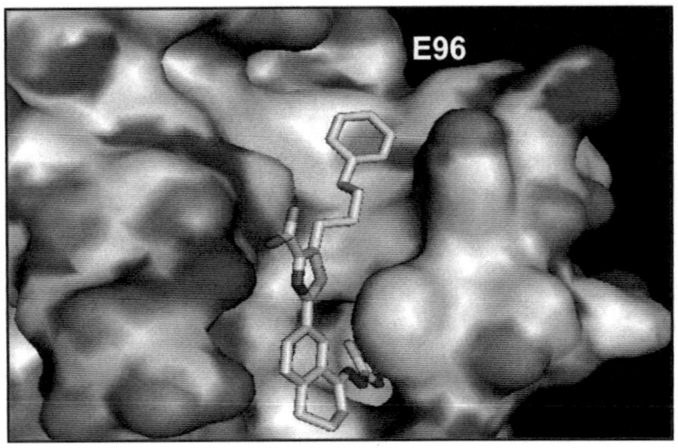

5

Bcl-x_L:11 nM (IC50)

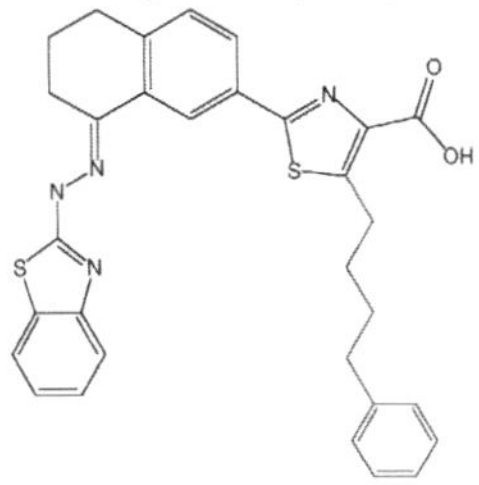

Peter Czabotar, Peter Colman (WEHI)

ABT-737 has an Intriguing interaction with E96

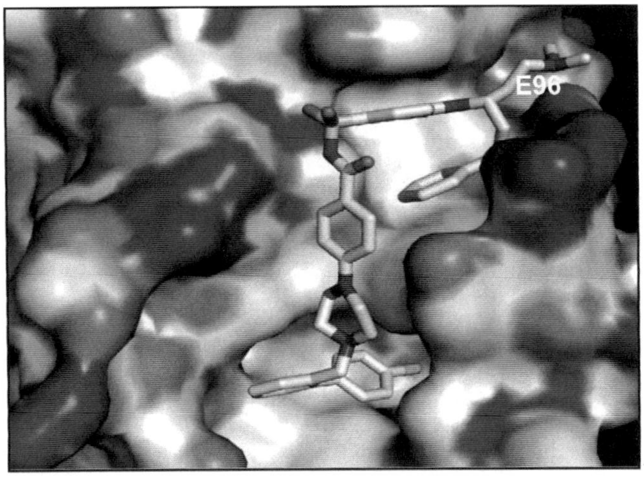

ABT-737

Bcl-x_L: <1 nM (IC50)

Bcl-2: <1 nM (IC50)

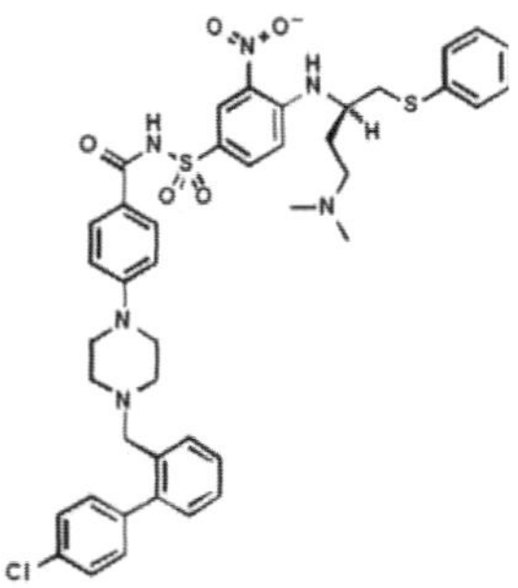

Peter Czabotar, Peter Colman (WEHI)

Addition of a Benzyl Amine Moiety to **5** markedly improves potency

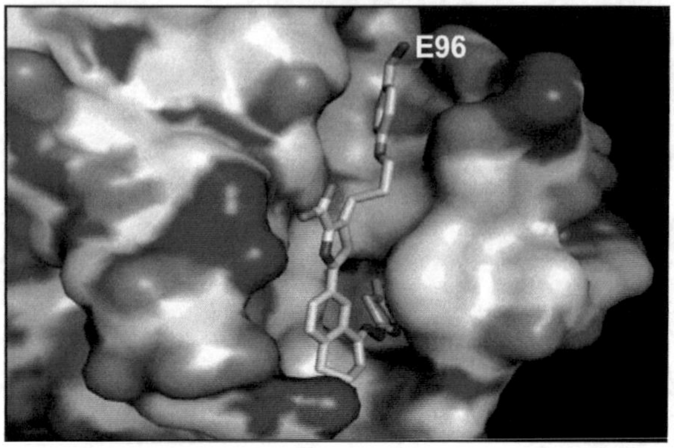

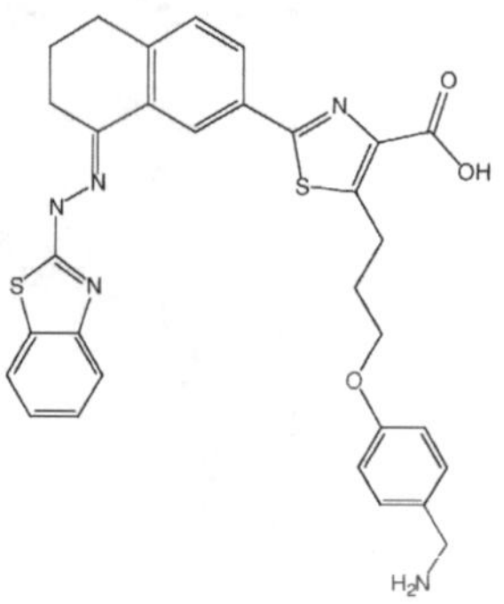

6

Compound **6** maintains excellent Bcl-X_L selectivity

Bcl-x_L: <1 nM (IC50)
Bcl-2: > 1000 nM (IC50)
MEFs: 70 nM (IC50, 10% serum, Mcl-1 knockout)
H146 (lymphoid cell lines) : 960 nM (EC50, 10% serum)

Peter Czabotar, Peter Colman (WEHI)

Bcl-x_L antagonism drives ABT-737 synergy with docetaxel on SKOV3 ovarian cancer line

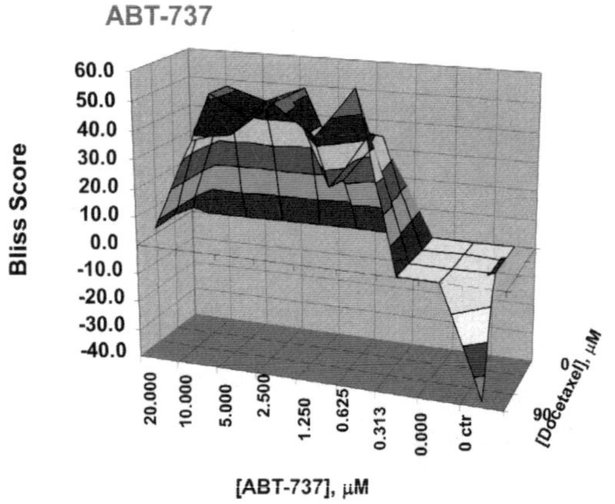

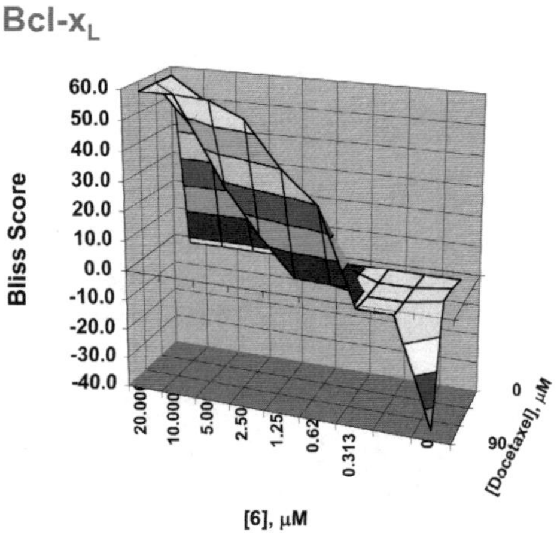

- Both ABT-737 and **6**, a Bcl-x_L-selective antagonist, enhance docetaxel cytotoxicity at low uM to sub-uM concentrations

Christin Tse, Abbvie Laboratories

Bcl-x_L antagonism drives ABT-737 synergy with docetaxel on BxPC3 pancreatic cancer line

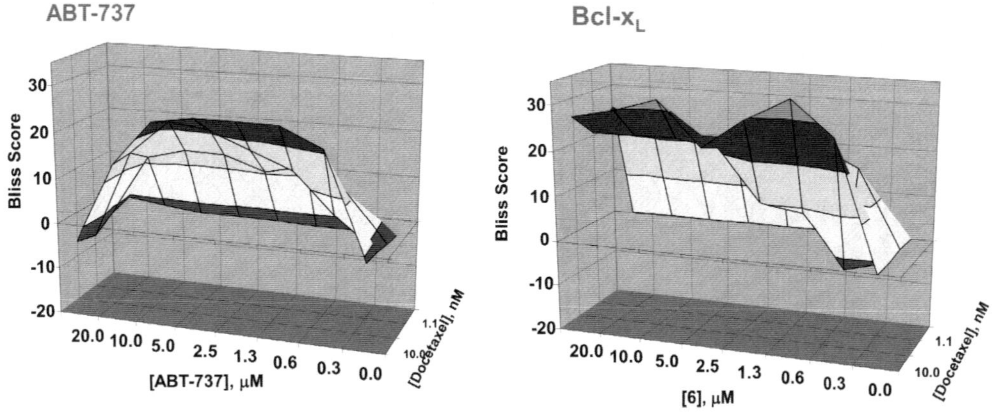

- Both ABT-737 and **6**, enhance docetaxel cytotoxicity at low uM to sub-uM concentrations

Christin Tse, Abbvie Laboratories

The Abbvie group found a key to Bcl-2 selectivity

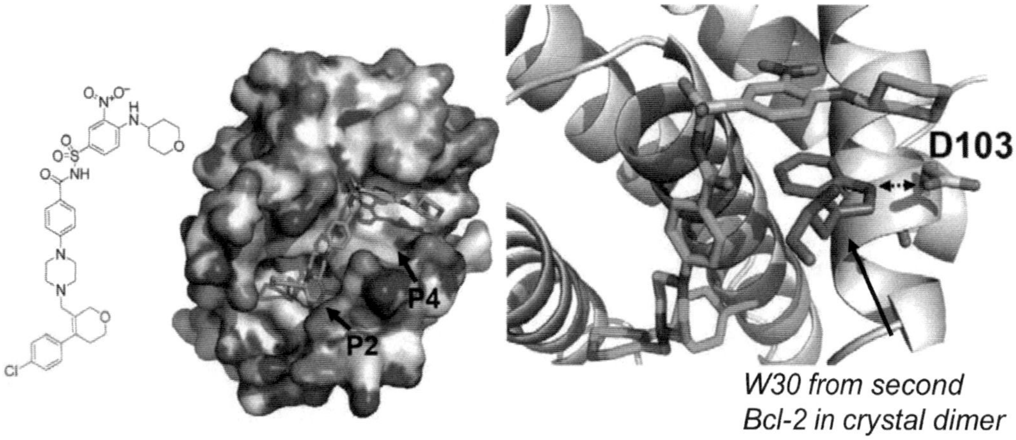

A. Souers et al. **Nature Medicine**, *In press*

Azaindole was incorporated to fill P4 and interact with D103

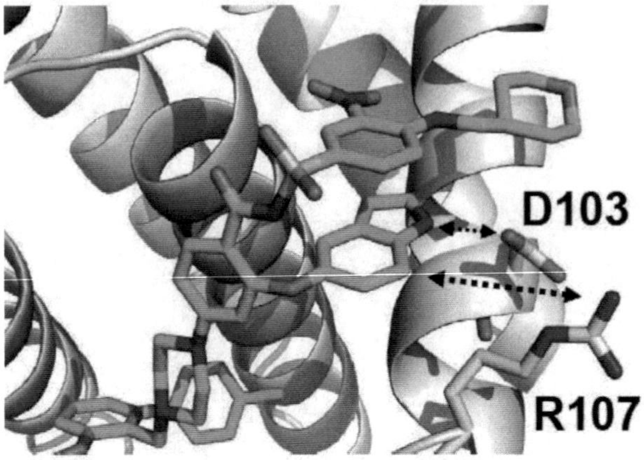

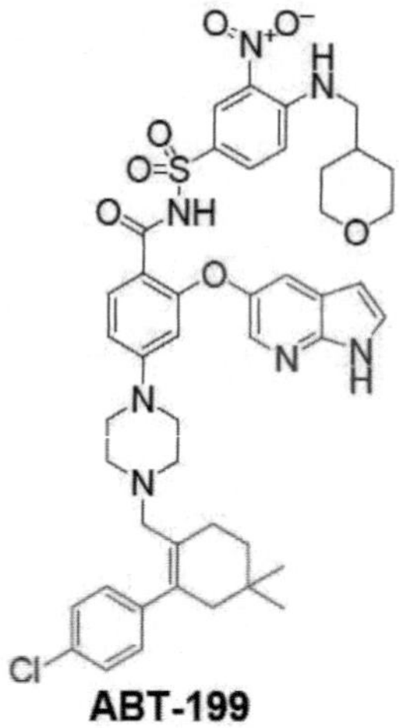

ABT-199

ABT-199 binds selectively to Bcl-2

ABT-199; Bcl-2: Bim = 1 nM, Bcl-xL:Bcl-xS=2 µM
ABT-263; Bcl-2: Bim =36 nM, Bcl-xL: Bcl-xS =38 nM

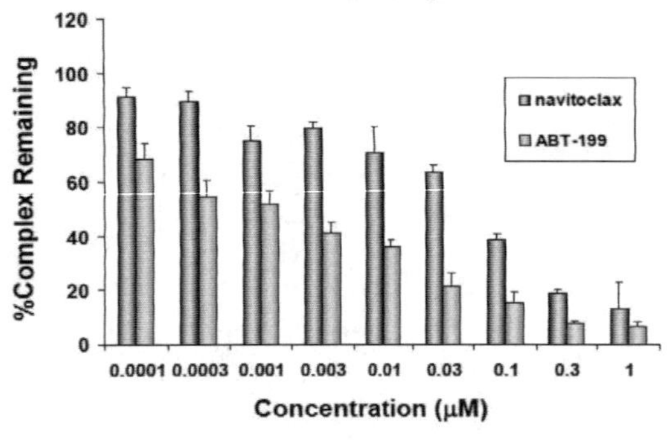

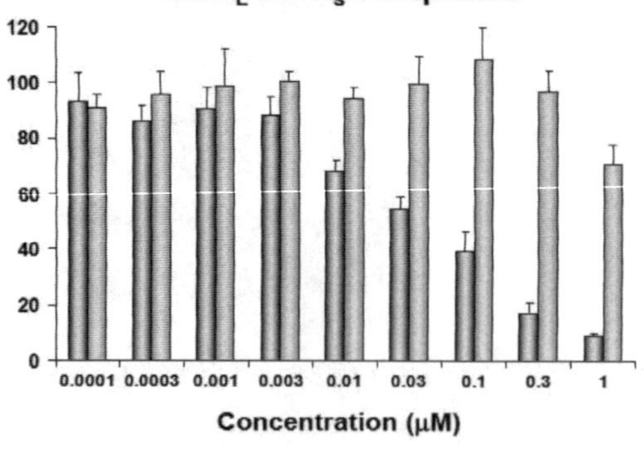

High Bcl-2 levels predict Sensitivity to ABT-199 in Non-Hodgkin's Lymphoma cell lines.

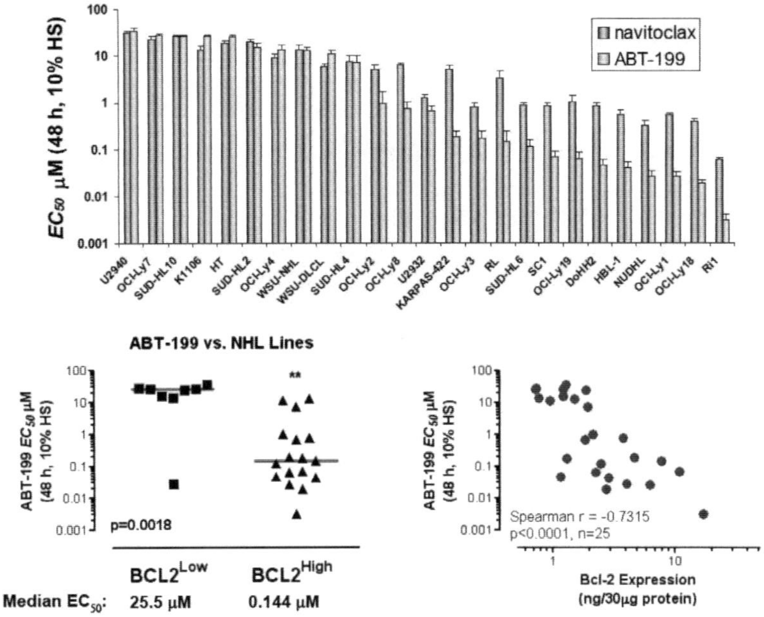

ABT-199 shows efficacy against primary patient-derived CLL cells and human patients.

Patient Blood Samples

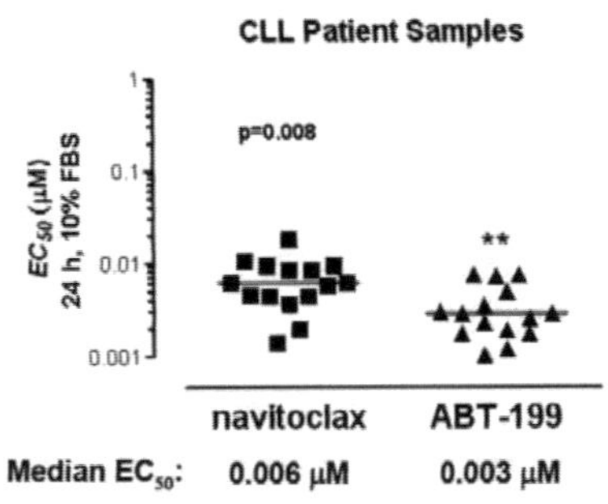

Patient Blood Count at 24 h (N=3)

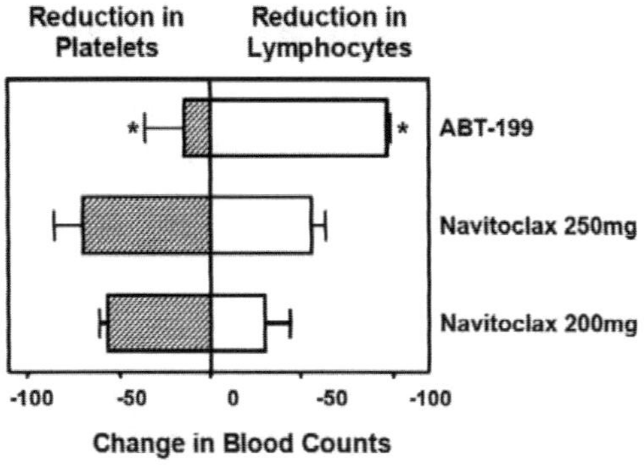

A. Souers et al. **Nature Medicine,** *In press*
A. Roberts et al **J. Clin. Uncol.** *(2012) 30,488-96*

Conclusions

Compounds can be developed that :

•Demonstrate sub-nanomolar potency against Bcl-X_L

•Show 1000-fold selectivity for Bcl-X_L over Bcl-2

•Show ~100-fold selectivity for Bcl-2 over Bcl-X_L

•Can be very specific for lymphomas (Bcl-2 selective)

•Evidence that Bcl-X_L antagonism drives synergy of Bcl-2 family antagonists with docetaxel in Ovarian and Pancreatic cancer cell lines. (Bcl-XL selective)

•Should prove useful for elucidating the specific role of Bcl-X_L and Bcl-2 in cancer.

•Bcl-2 selective compound ABT-199 shows efficacy against CLL in human patients.

Summary

- Apoptosis can be modulated using small molecule regulators of PPIs.
- Large orally viable molecules are possible that show high potency and selectivity for specific protein interfaces
- Small molecules that interfere with PPIs have been found by peptidomimetics, fragment screen and HTS.
- Hard work but potentially very fruitful .

Acknowledgements: IAP Antagonists

- **EDB**

  WayneFairbrother
  Kerry Zobel
  Heidi Wallweber
  Matt Franklin
  Domagoj Vucic
  Sarah Wayson
  Eugene Varfolomeev
  Matt Franklin

- **Pharmaceutics**
  Minli Xie
  Priscilla Mantik
  Tina Leung
  Linda Bao

- **Pathology**
  Jiping Zha

- **Assays**
  Jean-Philippe Stephan
  Stacy Frankovitz
  Sally Fischer
  Maureen Beresini
  Linda Elliott
  Mark Lackner
  Klaus Wagner
  David Dornan

- **DMPK/In Vivo Groups**

  Helen Chan
  Harvey Wong
  Jason Halladay
  Matt Baumgardner
  Bilin Chou
  Cyrus Khojasteh
  Hank La
  Fran Lai
  Ronitte Libedinsky
  Emile Plise
  Young Shin
  Savita Ubhayakar
  Ran Wu Hwang
  Susan Wong

- **Translational Oncology**
  Lesley Murray
  Karl Doerner
  Janeko Bower
  Sandy Adams
  Angela McNutt
  Beth Blackwood
  Kristina West
  Klara Totpal

- **Medicinal Chemistry**

  John Flygare
  Fred Cohen
  Lewis Gazzard
  Mike Koehler
  Bainian Feng
  Chudi Ndubaku
  Lan Wang
  Jeff Tom
  Stephen Keteltas
  Cuong Ly
  Melisa Wong
  Joy Middleton
  Kevin Lau
  Philippe Bergeron
  Zhaoyang Wen
  Cliff Quan
  Mark Reynolds
  Nicole Collins
  Vickie Tsui
  Jim Marsters
  Alan Olivero

  Vishva Dixit
  Avi Ashkenazi
  John Prescott

Acknowledgements: Bcl-2 Family Antagonists

WEHI

Guillaume Lessene
Keith Watson
Jerry Adams
Suzanne Cory
Peter Colman
Peter Czabotar
David Huang
Jackie Lee
Seong Lin Khaw
Kylie Mason
Andrew Roberts
John Seymour

Genentech

Kerry Zobel
Heather Maecker
Wayne Fairbrother
John Flygare
Sarah Hymowitz
Deepak Sampath

Abbvie

Andrew Souers
Steven Elmore
Joel Leverson
Erwin Boghaert
Saul Rosenberg
Rod Humerickhouse
Christin Tse
Scott Ackler
Nathanial Catron
Jun Chen
Brian Dayton
Hong Ding
Sari Enschede
Sha Jin
Kennan Marsh
Michael Mitten
Paul Nimmer
Kennan Marsh
Lloyd Lam
Morey Smith
Gerard Sullivan
Stephen Tahir
Michael Wendt
Yu Xiao
John Xue
Haichao Zhang
Peter Kovar

Supplemental slides

Killing by BH3-only proteins requires Bax or Bak

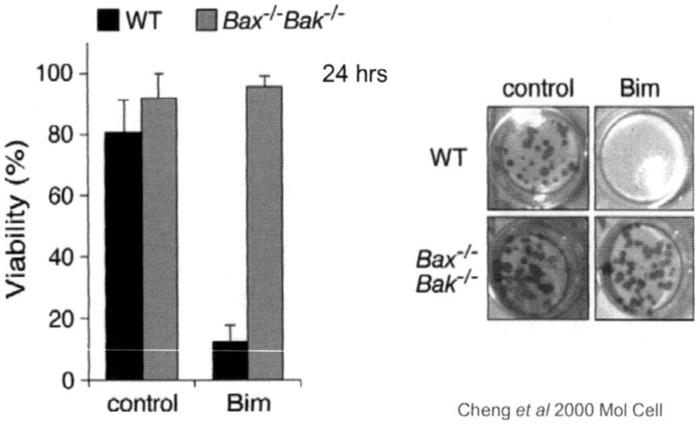

Cheng *et al* 2000 Mol Cell
Zong *et al* 2000 Genes & Dev.
van Delft *et al* 2006 Cancer Cell

The action of a true BH3 mimetic must also require Bax and/or Bak.

Most reported putative BH3 mimetics kill non-specifically

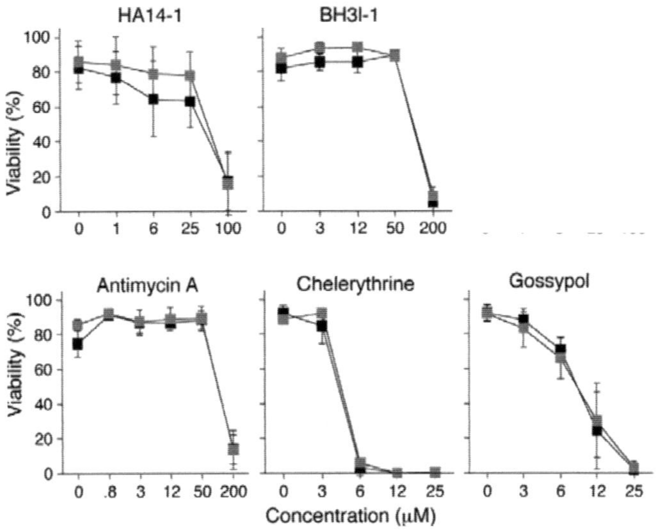

Weak affinity for Bcl-2-like proteins compared to BH3-only proteins.

Structural improvements may enhance specificity.

Abbvie's BH3 mimetic ABT-737

Structure-based design to target Bcl-x_L

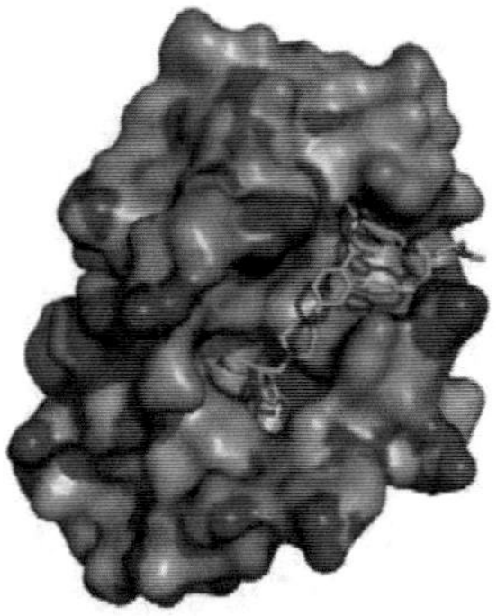

Lee *et al* 2007 Cell Death Diff

Promising single agent activity against certain tumors

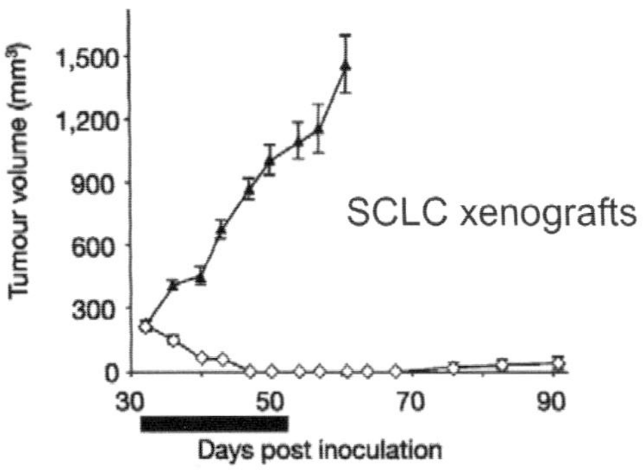

Oltersdorf *et al* 2005 Nature

sys²bio
www.sysbio.se

Yeast as a Platform Cell Factory in Future Biorefineries

Jens Nielsen

Systems & Synthetic Biology

Department of Chemical and Biological Engineering, Chalmers University of Technology, Sweden

Novo Nordisk Foundation Center for Biosustainability, Technical University of Denmark, Denmark

Science for Life Laboratory, Royal Institute of Technology, Sweden

CHALMERS

The Biorefinery Concept

sys²bio
www.sysbio.se

Metabolic Engineering of Cell Factories allows to develop novel cell factories for production of advanced biofuels

Existing infrastructures can be upgraded to produce new biofuels and biochemicals

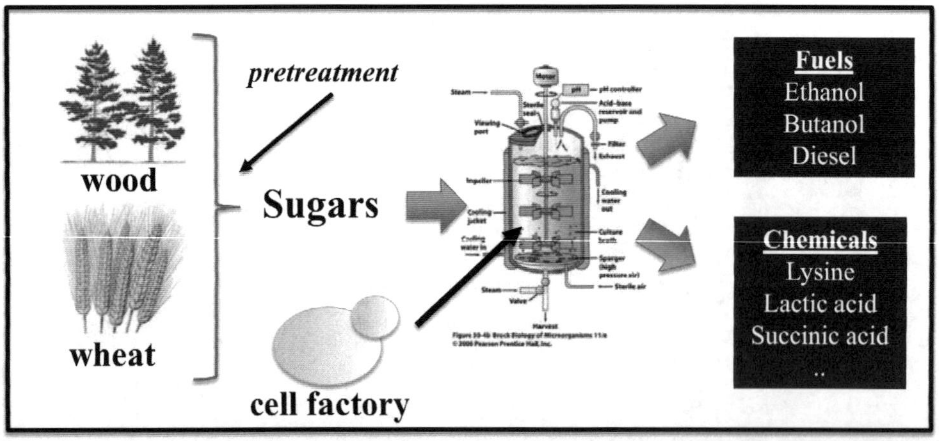

Cell. Mol. Life Sci. (2012) **16**:2671-2670

CHALMERS 2

Yeast as a Platform Cell Factory

sys²bio www.sysbio.se

Many advantages:

- Extremely well-characterized
- Many online databases with information on genome, as well as different omics data
- Genetically tractable
- GRAS (Generally Regarded as Safe)
- **Robust industrial organism**

A widely used cell factory

- Bioethanol
- Baker's yeast
- Wine & Beer
- Resveratrol
- Insulin precursors
- Vaccines (HPV, Hepatatis)

Ongoing developments of novel cell factories:

- **Fuels** (butanol, biodiesel)
- **Commodity chemicals** (malate, succinate, 3-OH propionic acid)
- **Fine chemicals** (isoprenoids**)**
- **Food ingredients** (PUFAs)
- **Protein drugs**

FEMS Yeast Research (2008) **8**:122-131
FEMS Yeast Research (2012) **12**:228-248

CHALMERS 3

Metabolic Engineering of Yeast

sys²bio www.sysbio.se

Our objective is to establish an extensive technology base for wider use of yeast as platform cell factory and demonstrate its use for development of diverse products

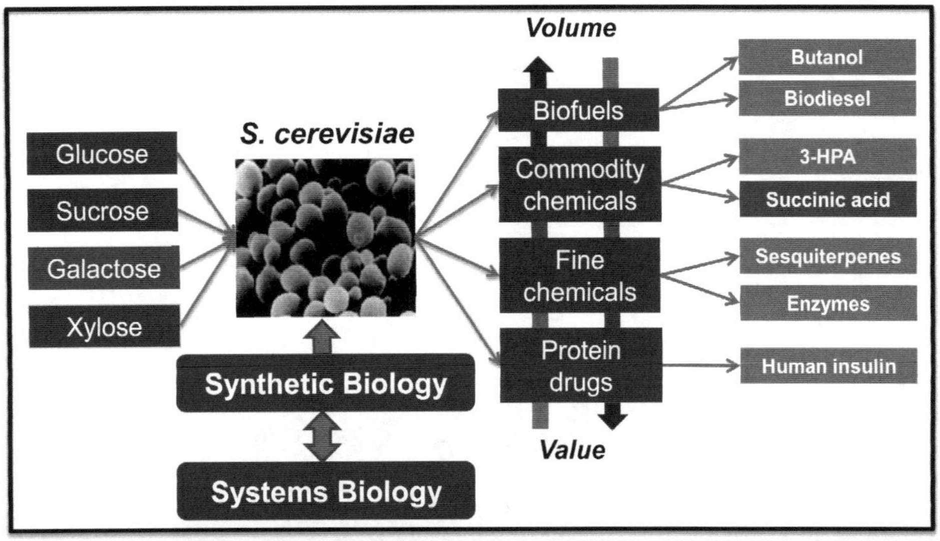

Cell. Mol. Life Sci. (2012) **16**:2671-2670

CHALMERS 4

Metabolic Engineering

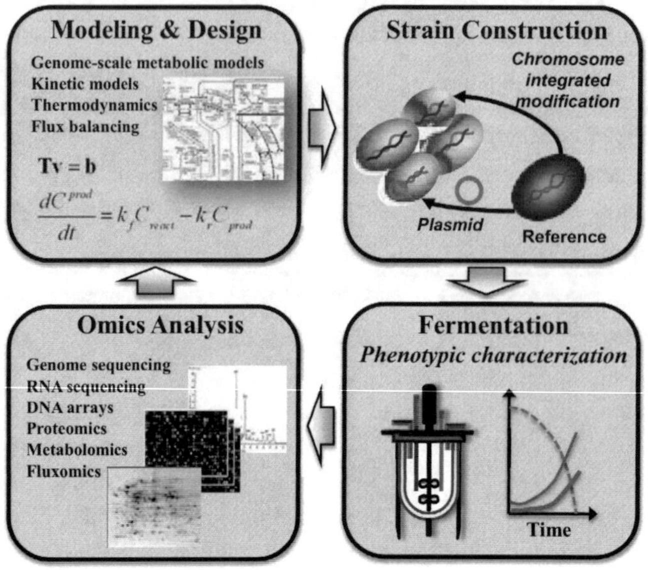

Appl. Microbiol. Biotechnol. (2001) **55**:263-283
Metabolic Eng. (2003) **5**:56-69; *Bioengineered* **4**:4

5

Metabolic Engineering

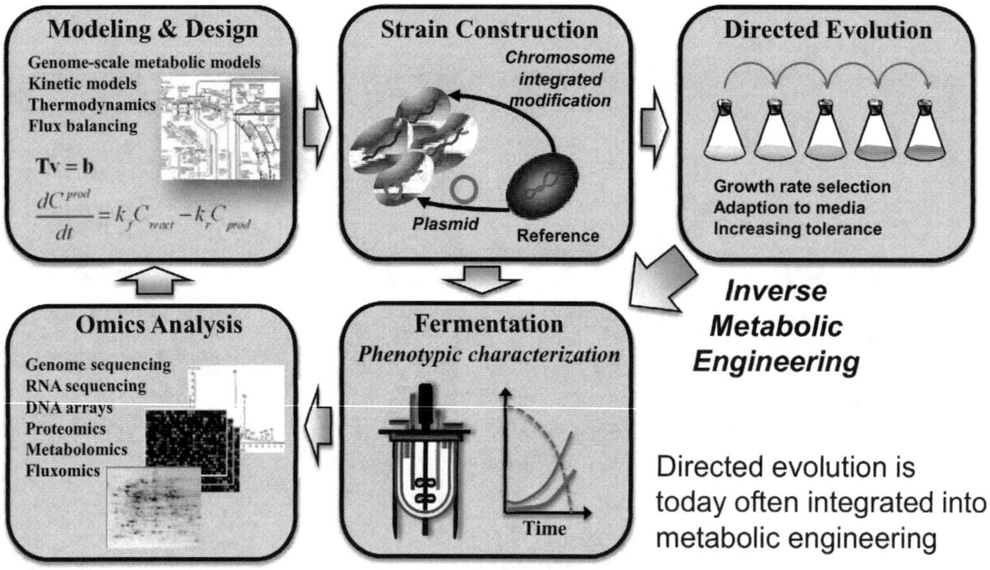

Directed evolution is today often integrated into metabolic engineering

Metabolic Eng. (2004) **6**:204-211

6

Inverse Metabolic Engineering

Through analysis of strains with different phenotypes new metabolic engineering targets may be identified

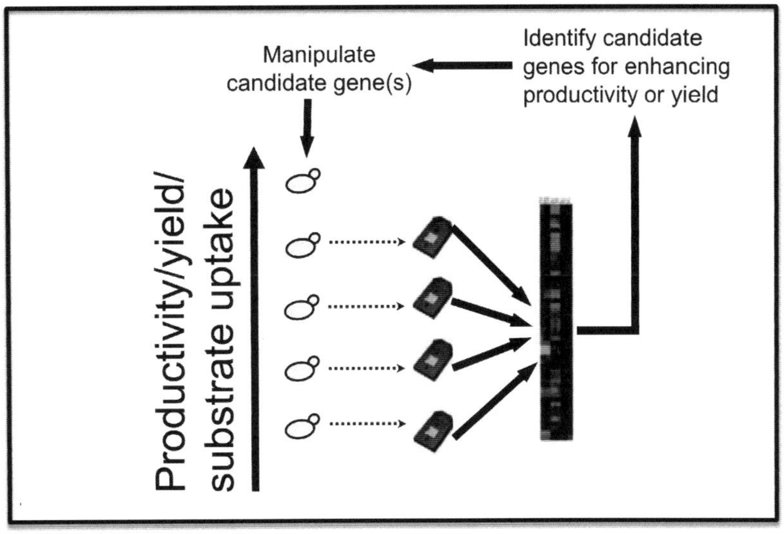

Metabolic Eng. (2004) **6**:204-211

Metabolic Engineering

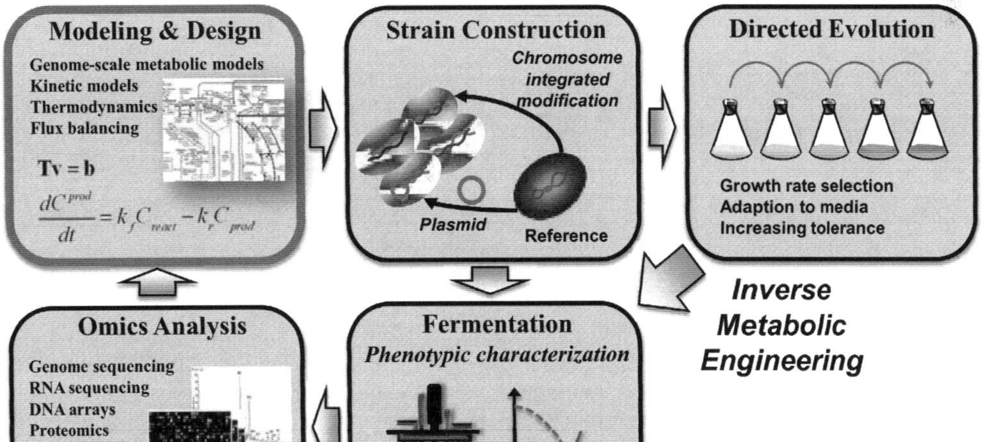

Directed evolution is today often integrated into metabolic engineering

Metabolic Eng. (2004) **6**:204-211

Genome-Scale Metabolic Models

sys²bio www.sysbio.se

Genome-scale metabolic models (GEMs) provide gene-protein-reaction connections and hereby allow for context dependent analysis

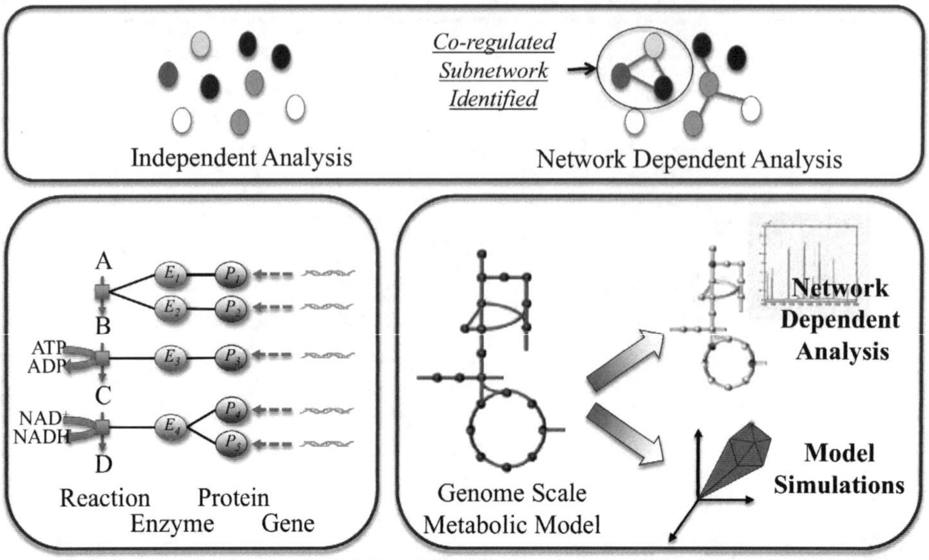

PNAS (2005) **102**:2685-2689; *Mol. Sys. Biol.* (2006) **2**:50
Curr. Op. Microbiol. (2010) **13**:255-262; *Biotechnol. Adv.* (2012) **30**:979-988

CHALMERS 9

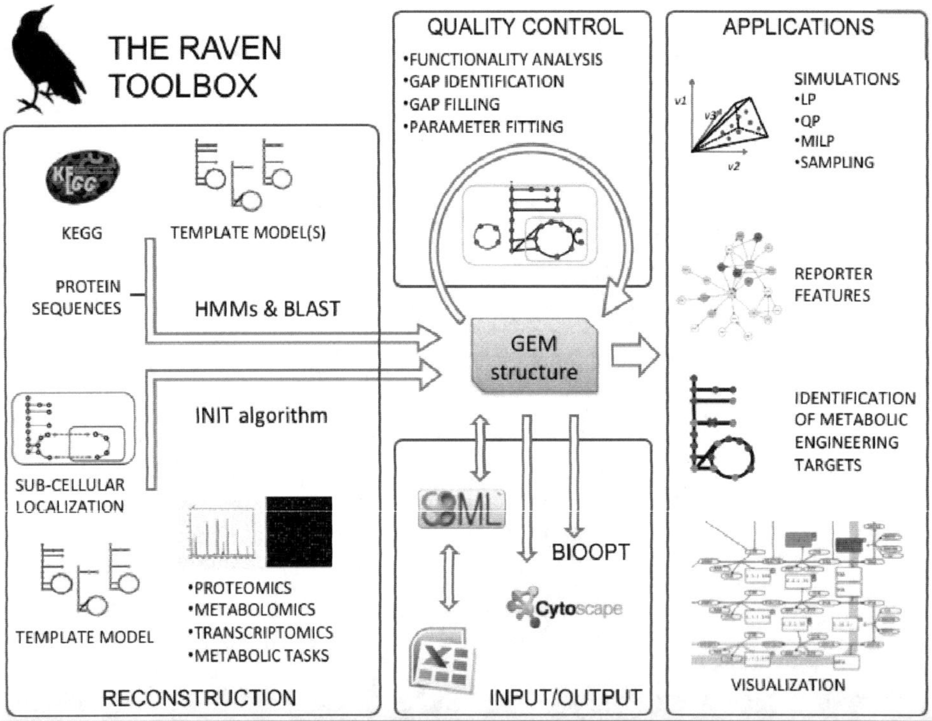

io bio.se

Submitted for publication

CHALMERS 10

BioMet Toolbox – Fungal Models

Organism	Model name	Metabolic genes	Unique metabolic reactions	Metabolites
S. cerevisiae	iFF708	708	840	555
	iIN800	800	1061	683
A. nidulans	iHD666	666	791	244
A. niger	iHD20	20	318	542
	iMA871	871	1195	792
A. oryzae	iWV1314	1314	1246	805
P. chrysogenum	iAL1008	882	1496	1012
P. pastoris	iLC915	915	1423	899
P. stipidis	iSS884	884	1332	922

Models and analysis tools (reporter algorithm) available at **www.sysbio.se**

NAR (2010) **38**:W144-W149
BMC Systems Biol. (2012) **6**:24

11

New Yeast Model (iTO980)

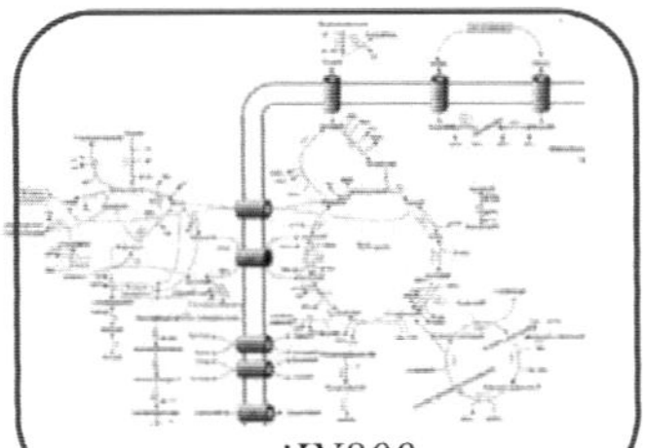

iIN800

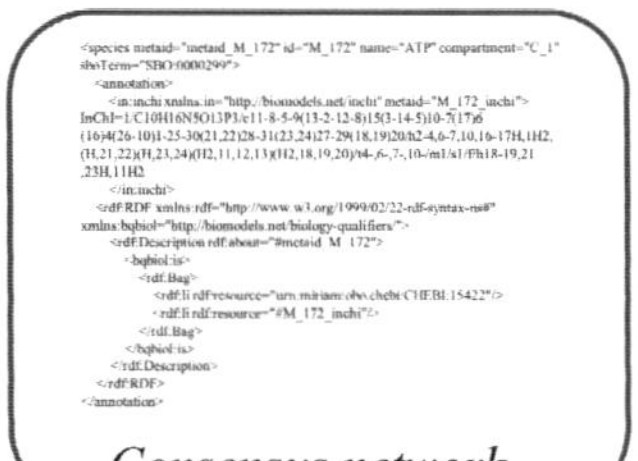

Consensus network

Draft model

Genome-scale model

1212 metabolites

1566 reactions

- 1083 cytosolic
- 256 mitochondrial
- 80 pero:

Nature Biotechnol. (2008) **26**:1155-1160; *Genome Res.* (2003) **13**:244-253
BMC Systems Biol.. (2008) **2**:7; submitted for publication

12

Identification of ME Targets

We developed the evolutionary algorithm **OptGene** for identification of targets for metabolic engineering

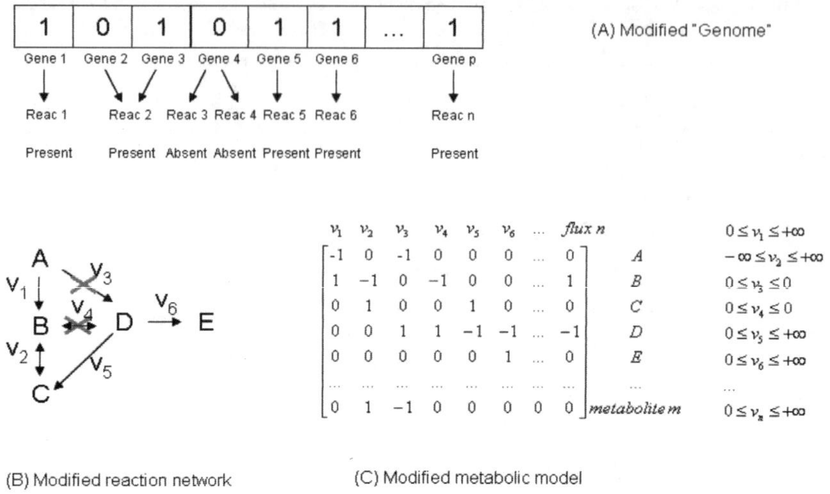

BMC Bioinformatics (2005) **6**:308

13

Succinate Production

OptGene was used to identify ME targets

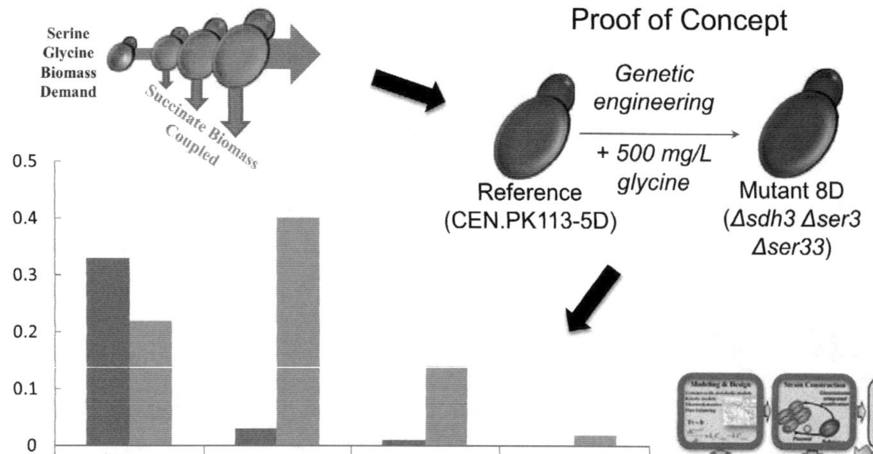

	Specific Growth Rate (1/h)	Maximum Titer (g/L)	Maximum Yield (g/g-biomass)	Maximum Yield (g/g-glucose)
REF	0.33	0.03	0.01	0.00
8D	0.22	0.40	0.14	0.02

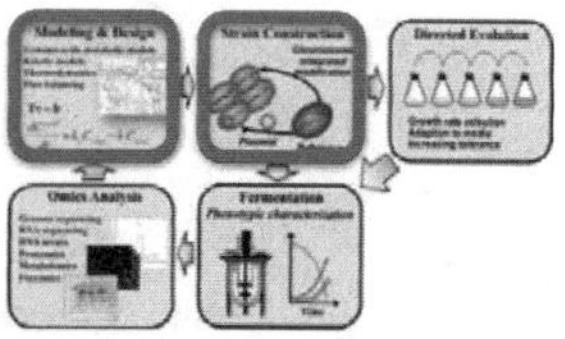

PLoS One (2013), *in press*

14

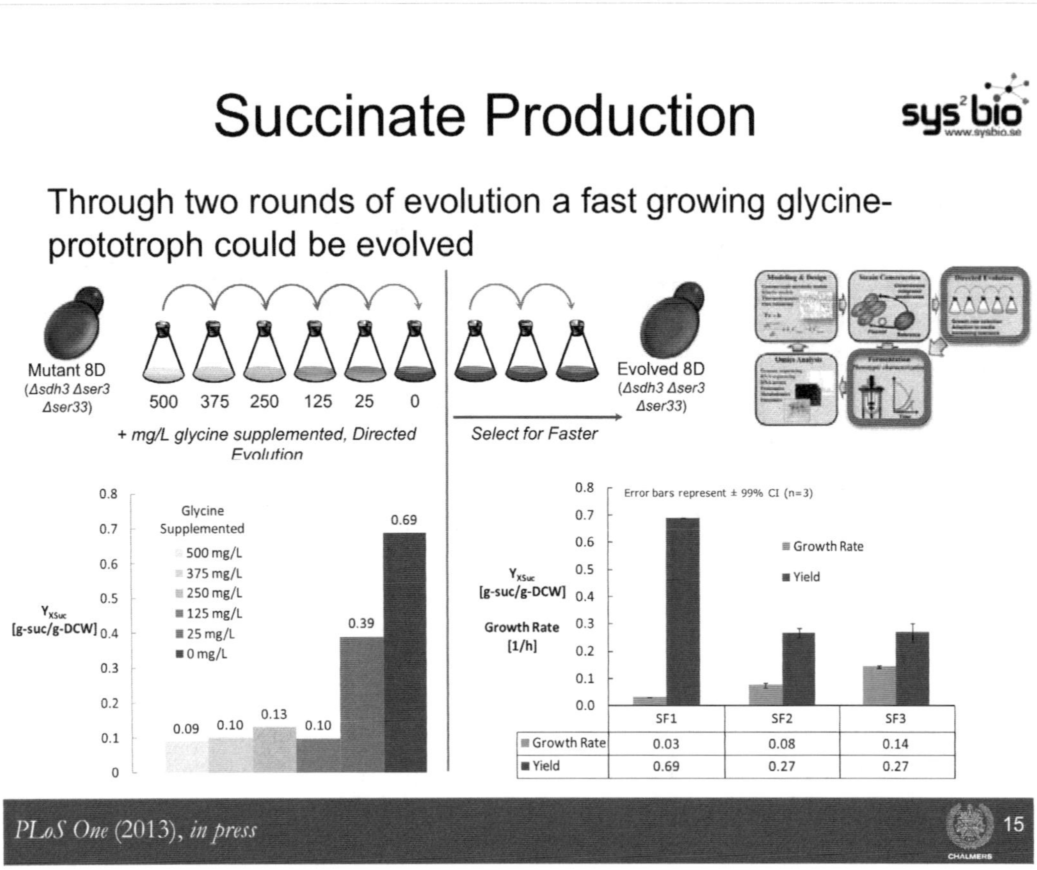

	SF1	SF2	SF3
Growth Rate	0.03	0.08	0.14
Yield	0.69	0.27	0.27

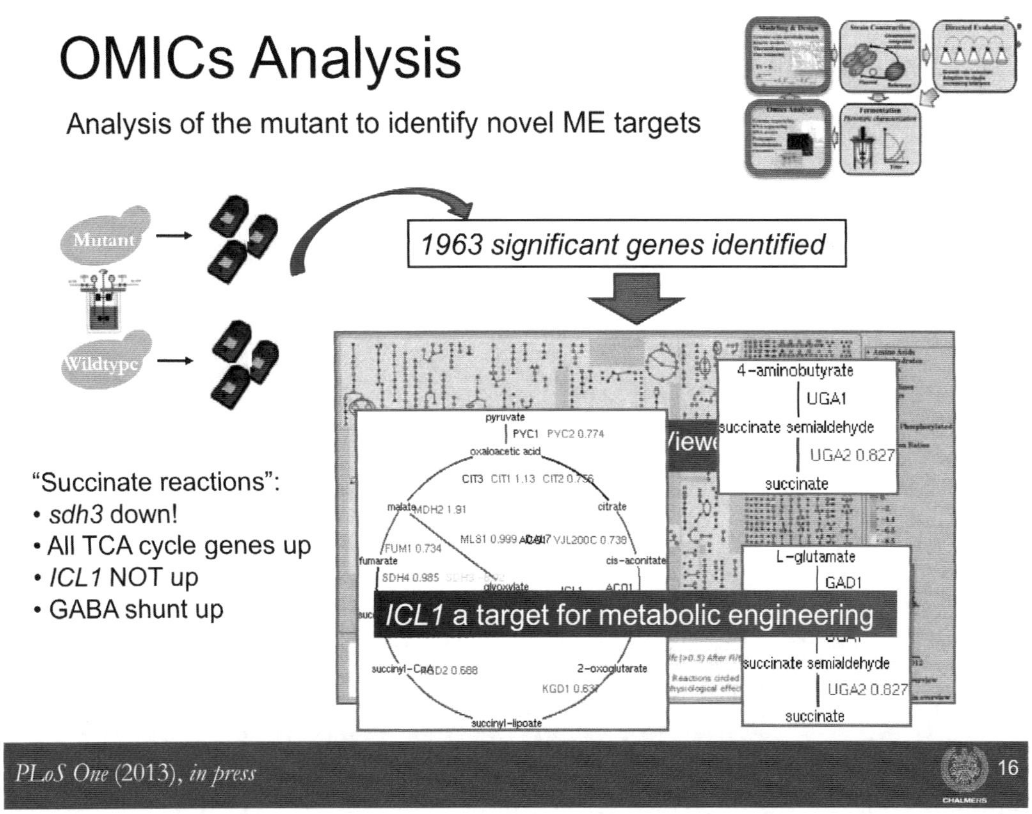

Over-Expression of *ICL1*

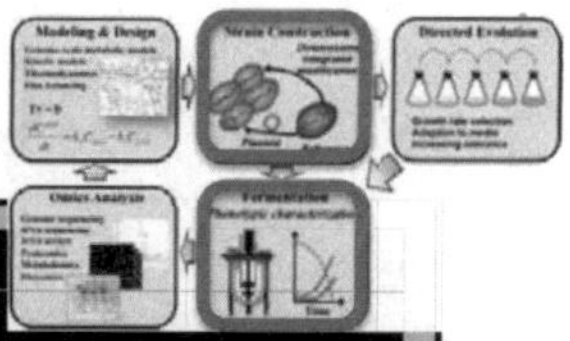

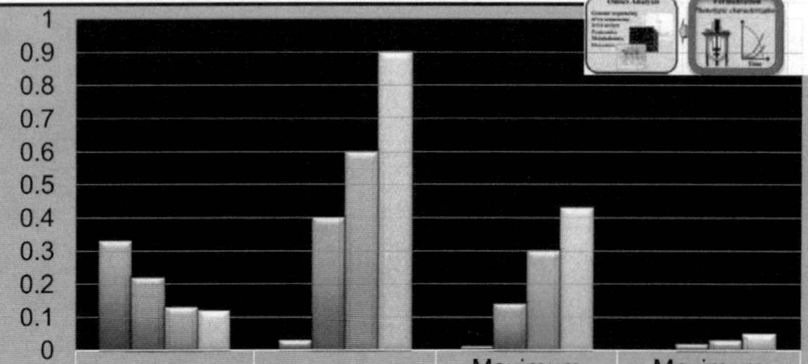

	Maximum Growth Rate	Maximum Titer (g/L)	Maximum Yield (g/g-biomass)	Maximum Yield (g/g-glucose)
REF	0.33	0.03	0.01	0
8D	0.22	0.4	0.14	0.02
8D Evolved	0.13	0.6	0.3	0.03
8D Evolved + pICL1	0.12	0.9	0.43	0.05

Model based design, adaptive evolution, and directed engineering resulted in a 50 improvement in titer and yield

PLoS One (2013), *in press*

17

CHALMERS

Galactose Metabolism

sys²bio
www.sysbio.se

Galactose is a dominant sugar in many hemi-celluloses

Galactose uptake is much slower than glucose uptake

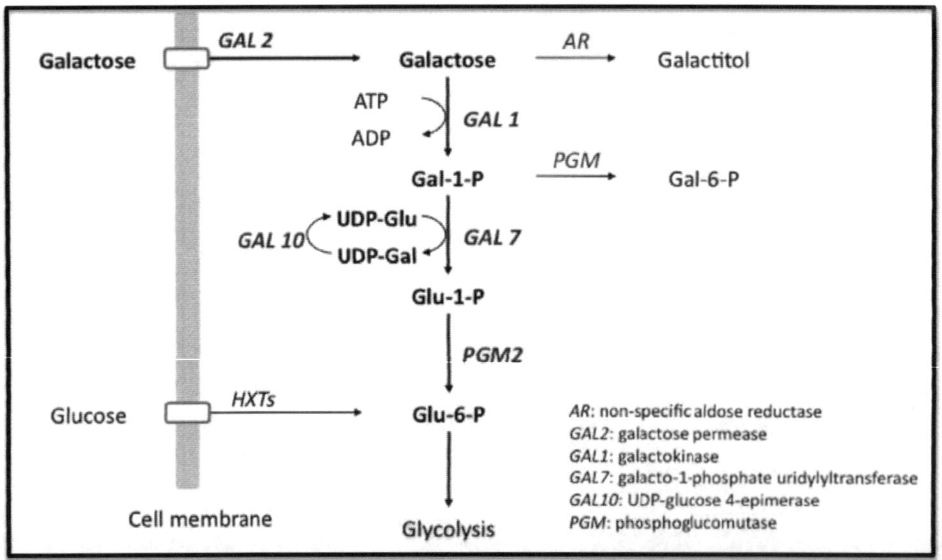

18

CHALMERS

ME of Galactose Metabolism

We over-expressed different combination of the catabolic genes in the Leloir pathway

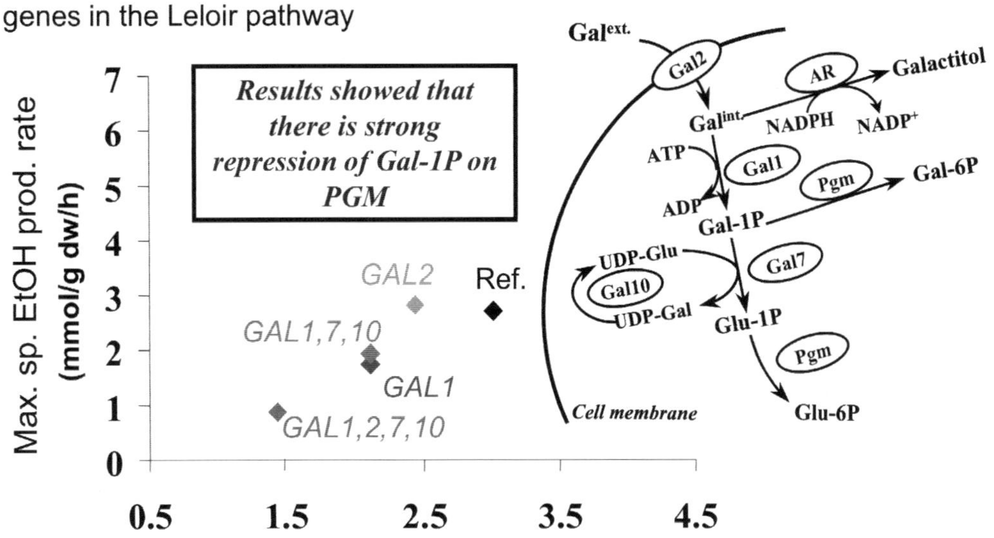

B&B (2008) **101**:317-326

19

Galactose Metabolism: Regulation

Galactose metabolism is tightly regulated

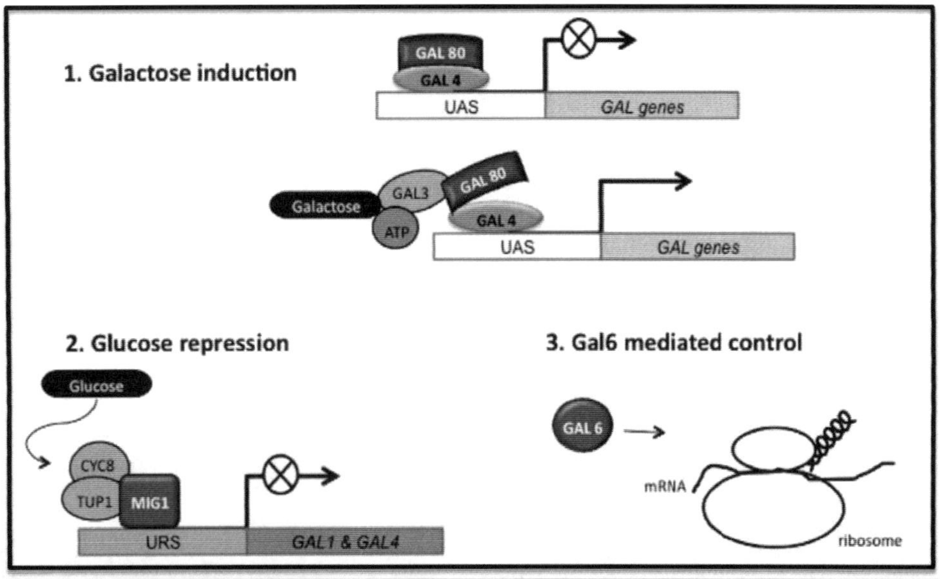

20

Galactose Metabolism: Regulation

Galactose metabolism is tightly regulated

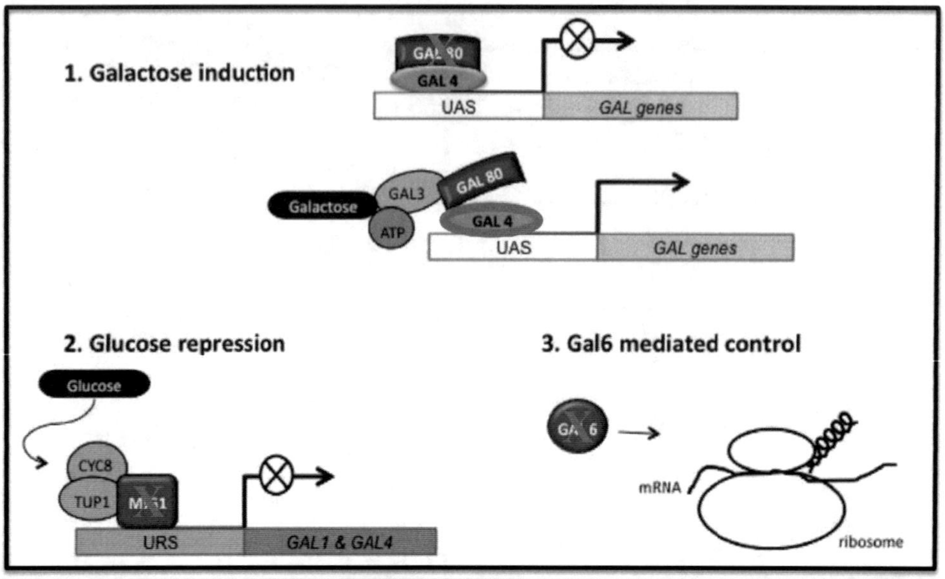

Nature Biotechnol. (2000) **18**:1283-1286

21

Engineering the Regulatory Network

Engineering of the regulatory network results in a 40% increase in galactose uptake

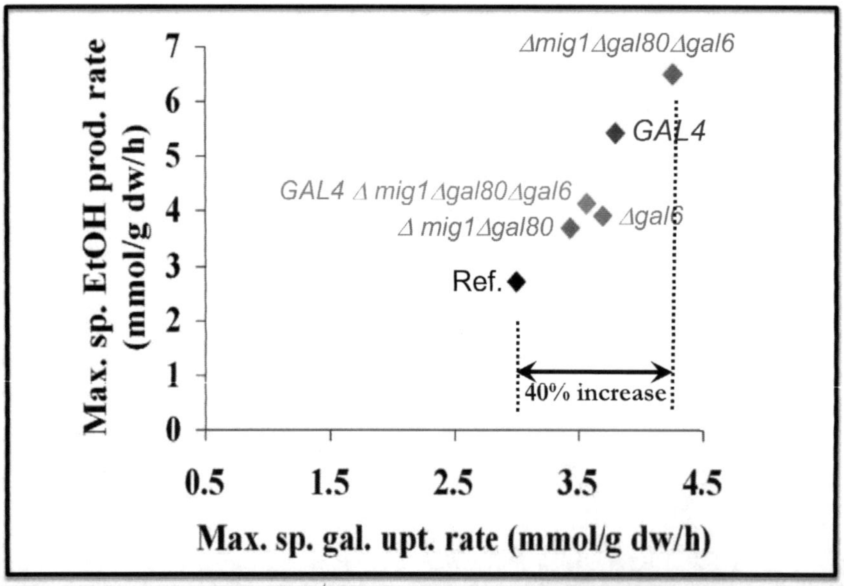

Nature Biotechnol. (2000) **18**:1283-1286

22

Engineering the Regulatory Network

Engineering of the regulatory network also results in glucose de-repression and consequently the uptake of glucose and galactose is occurring at the same time

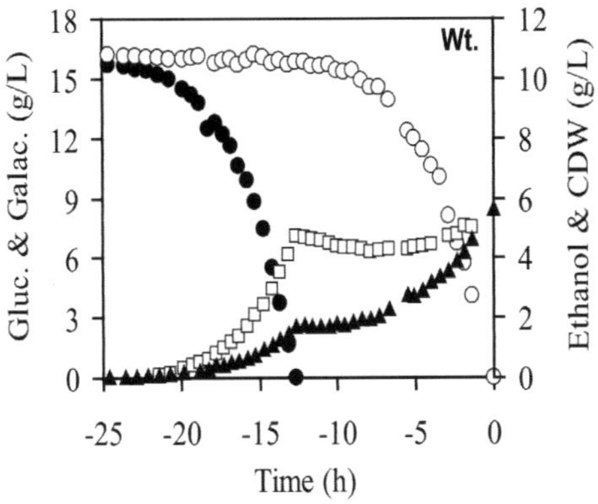

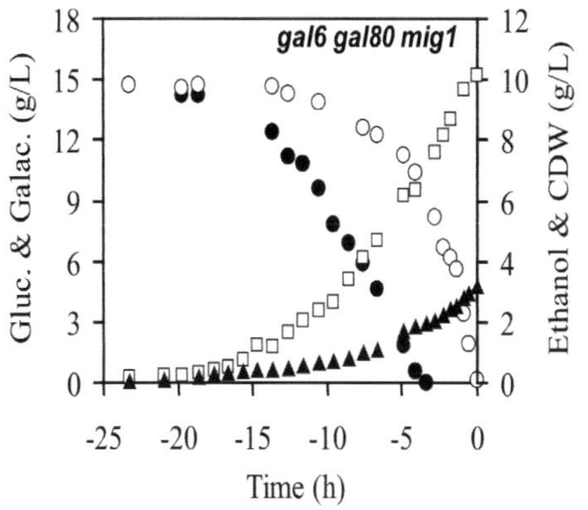

FEMS Yeast Research (2001) **1**:47-55

Inverse Metabolic Engineering

Transcription analysis of SO7 and SO16 identified Pgm2 as a potential target for improving galactose metabolism

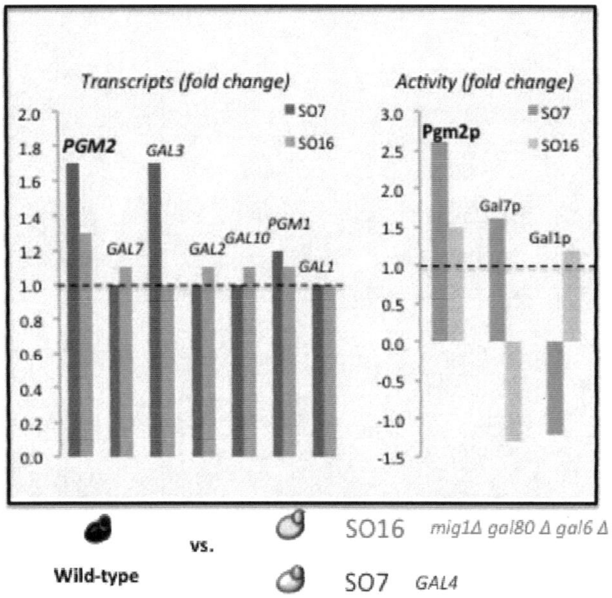

Appl. Environ. Microbiol. (2005) **71**:6465-6472

Over-Expression of *PGM2*

Over-expression of *PGM2* alone resulted in a 80% increase in galactose uptake rate

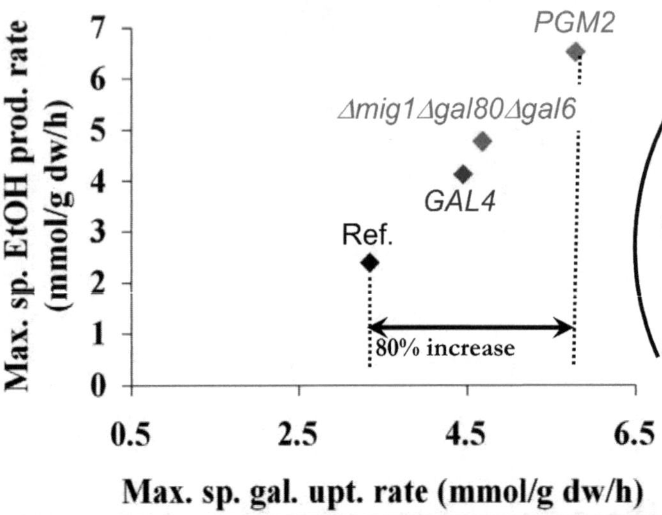

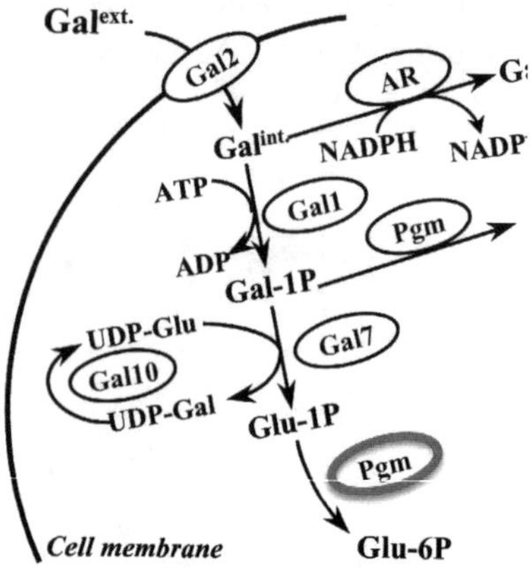

Appl. Environ. Microbiol. (2005) **71**:6465-6472

25

Adapative Evolution

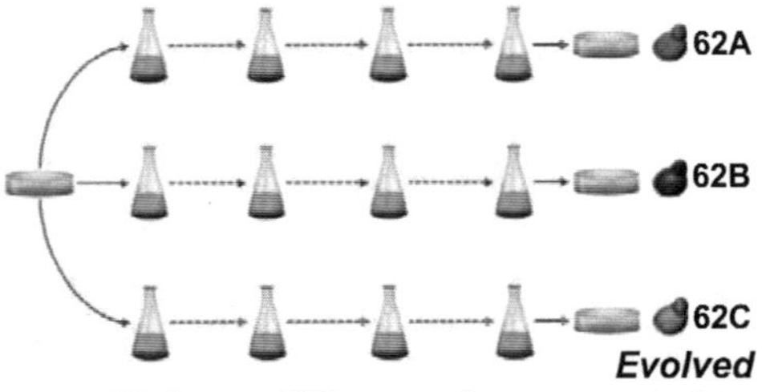

62 days, ~400 generation, mid-exponential phase transfers

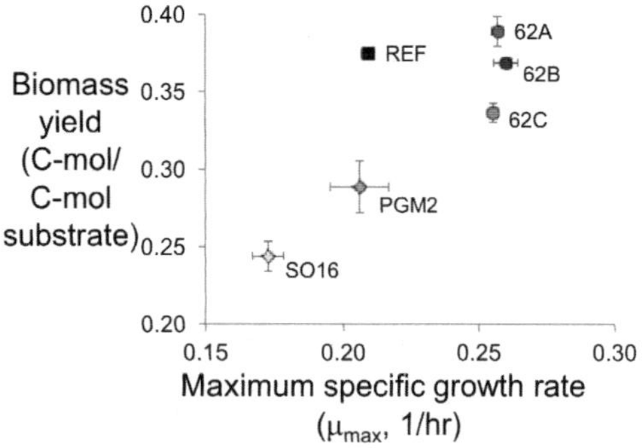

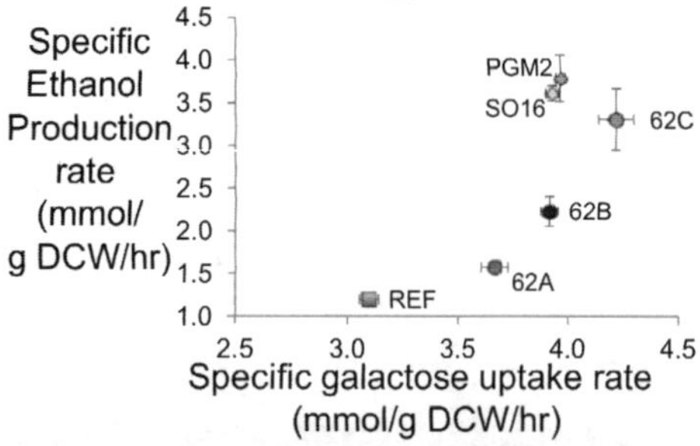

PNAS (2011) **108**:12179-12184

26

Adapative Evolution

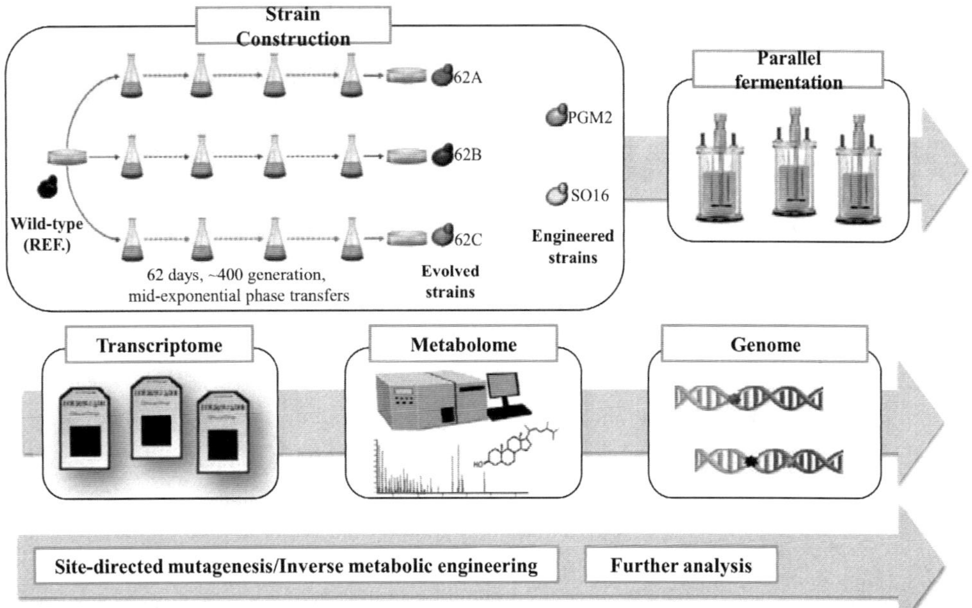

PNAS (2011) **108**:12179-12184 27

Transcriptome Analysis

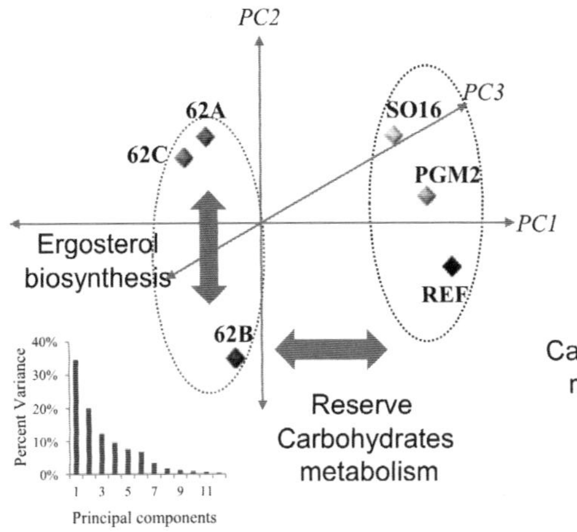

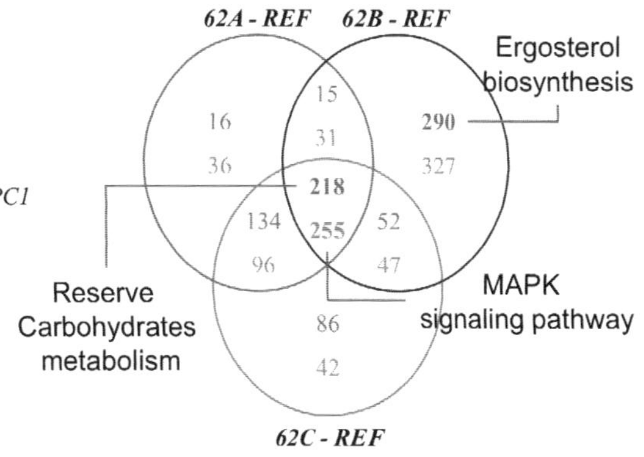

PNAS (2011) **108**:12179-12184 28

Metabolome Analysis

Evolved mutants differ from the engineered strains, and 62A and 62C cluster whereas 62B has a distinct metabolome profile

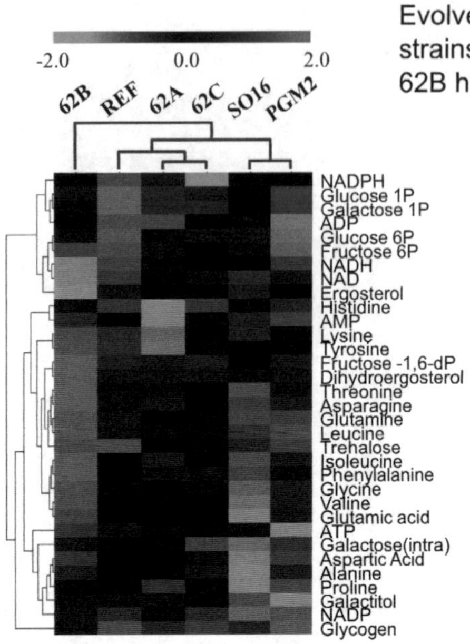

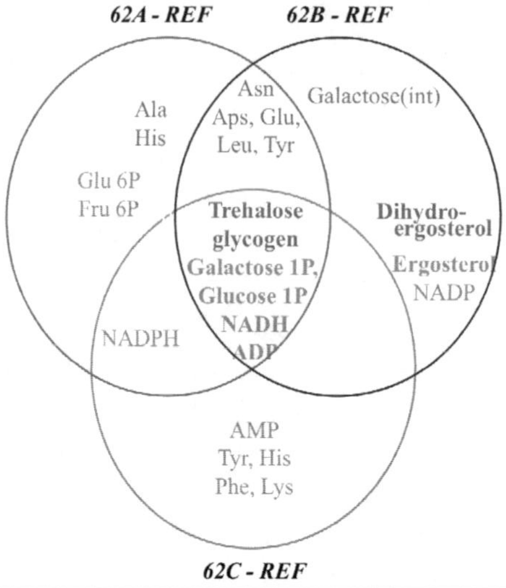

Integration of Transcriptome and Metabolome

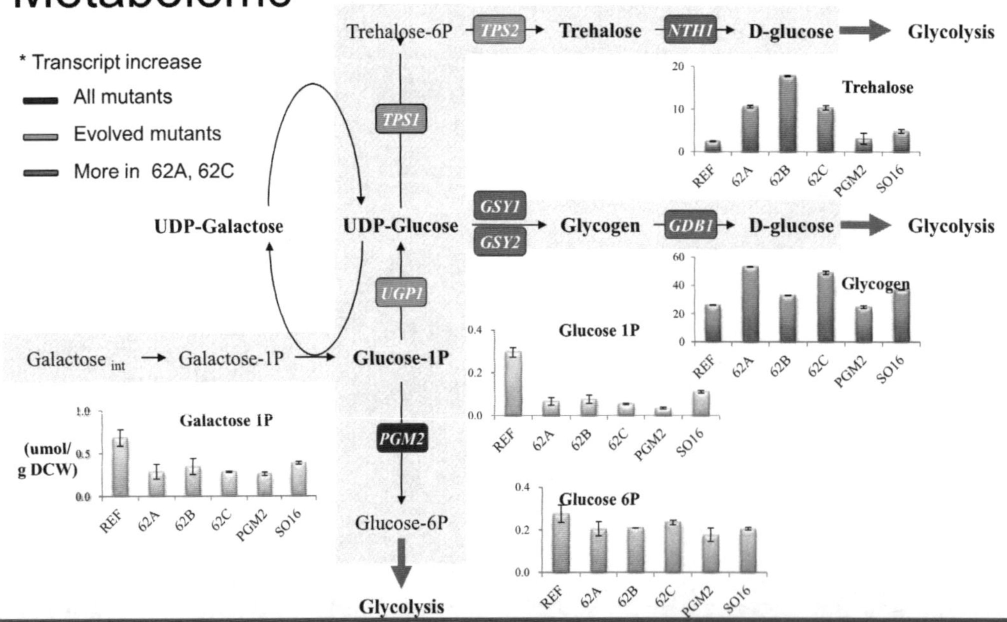

Genetic changes

	62A/Others[a]	62B/Others	62C/Others
Total number of mutations	47	347	43
Total number of SNVs	23	111	31
Coding region	7	31	12
Non coding region	16	80	19
Total number of Insertions and deletions	24	236	12
Coding region	0	12	3
Non coding region	24	224	9

[a] All other strains including the reference strain, CEN.PK113-7D and the other evolved strains

No mutations in galactose (*PGM2*) and storage carbohydrate metabolism

→ Different strategies from rational engineering

→ need to identify **"driving mutations" (beneficial mutations for galactose utilization)**

PNAS (2011) **108**:12179-12184

31

Genetic changes

Possible mutations driving changes of *PGM2*, storage carbohydrate and ergosterol metabolism

Strains		Genes/Proteins related to storage carbohydrate accumulation	References
62A	Coding (3)	*RAS2*[Gln^{77} -> Lys]	*Cannon JF et al. (1994)*
		OSH2[Arg^{1202} -> Cys]	*Wilson Wa et al. (2002)*
	Non-coding (2)	*MRPS5*[U][b], *SRB2*[D][a]	*Wilson Wa et al. (2002)*
62B	Coding (1)	*RAS2*[Asp^{112} -> Tyr]	*Cannon JF et al. (1994)*
	Non-coding (23)	*YAP3*[U], *MSR1*[U], *MRPL49*[D], *UBR2*[D], *SNF7*[U], *RPS0B*[D], *CSF1*[D], *LCB5*[D], *SEC22*[U], *DCS1*[U], *RPS16A*[U], *IMP1*[U], *PIN4*[D], *RSM18*[D], *FAR7*[U], *RTF1*[D], *DOC1*[D], YIL077C[D], YIL029C[D]	*Wilson Wa et al. (2002)*
		ATG13[D]	*Wang Z et al. (2001)*
		GLN3[D]	*Powers RW et al. (2006)*
		GIP2[D]	*Williams-HT et al. (2002)*
		GLC7[U]	*Cannon JF et al. (1994)*
62C	Coding	*CYR1* [Asp^{822} ->Asn]	*Francisco E (2000)*

Strains		Genes/Proteins related to ergosterol metabolism	References
62B	Coding (1)	*ERG5*[Arg^{370} -> Pro]	*Steven LK et al.(1995)*
	Non-coding (5)	*ERG1*[D], *ERG2*[U], *ERG25*[D], *ERG26*[D],*ERG28*[U]	

[a][D]: Downstream region
[b][U]: Upstream region

PNAS (2011) **108**:12179-12184

32

Genetic changes

Pathway conserved mutations

Strains	Mutations	Functions
62A	*RAS2* [Gln^{77} -> Lys]	Ras/PKA signaling pathway
62B	*RAS2* [Asp^{112} -> Tyr]	
62C	*CYR1* [Asp^{822} ->Asn]	

Unique mutation

Strains	Mutations	Functions
62B	*ERG5* [Arg^{370} -> Pro]	Ergosterol metabolism

Hypothesis

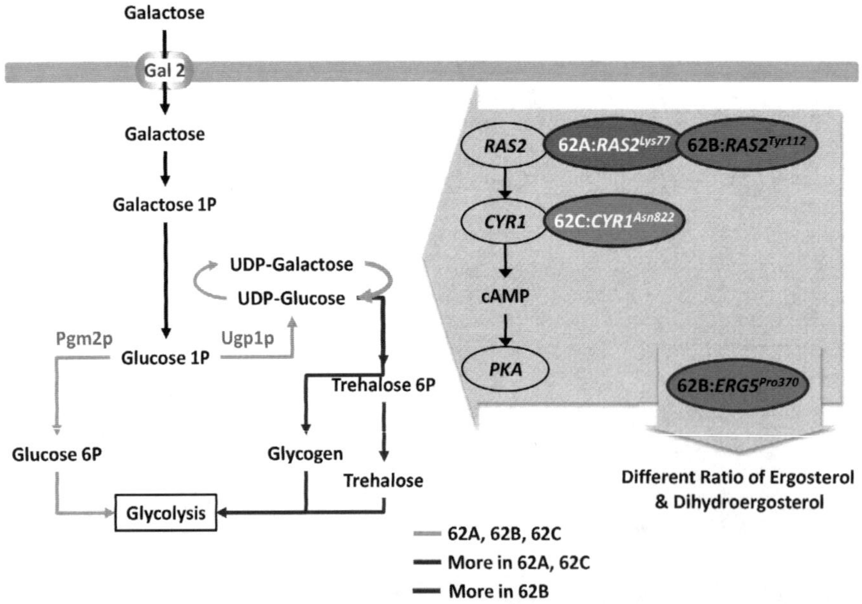

Validation

sys²bio www.sysbio.se

We evaluated two of the single nucleotide mutations identified in *RAS2* and one single nucleotide mutation in *ERG5*

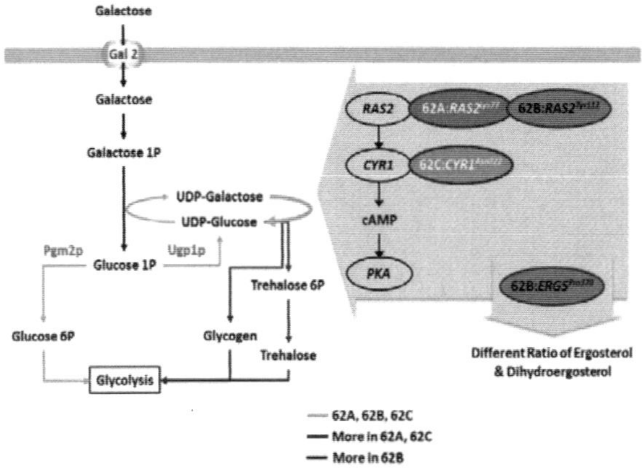

Improvement of specific galactose uptake rate (%)

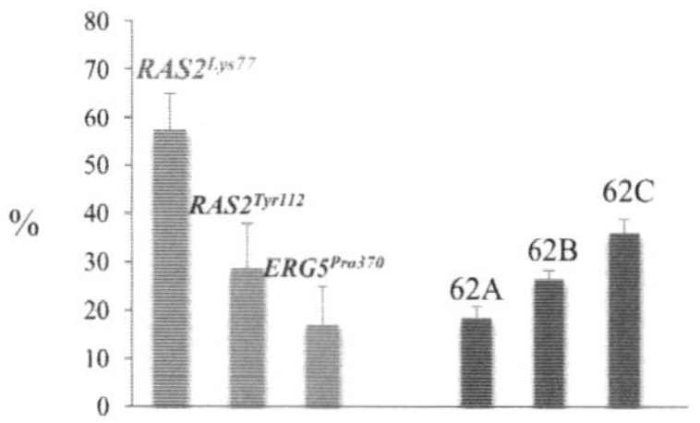

Improvement of maximum specific growth rate (%)

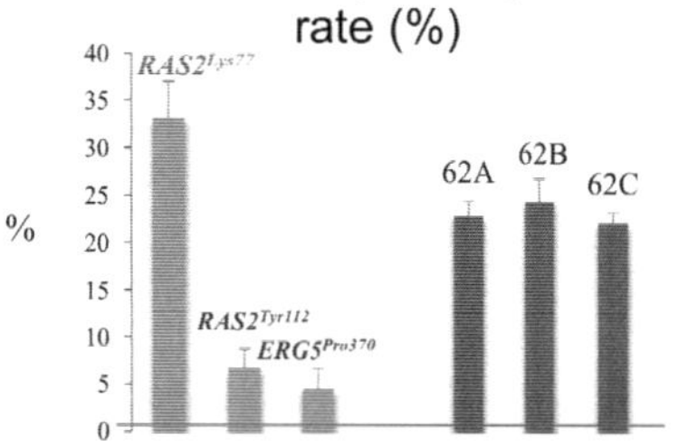

PNAS (2011) **108**:12179-12184

Inverse metabolic engineering

sys²bio www.sysbio.se

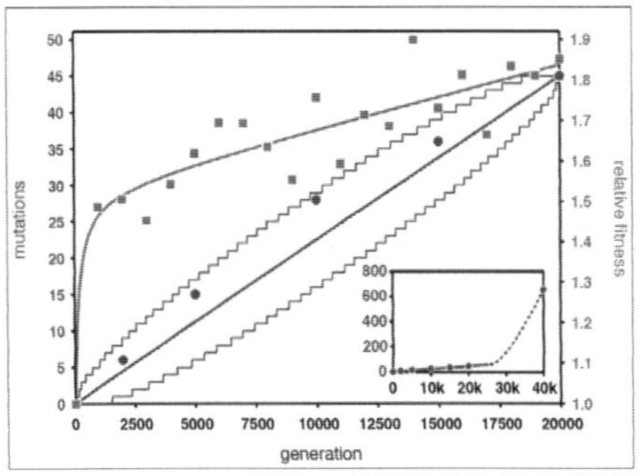

Nature (2009) ***461:****1243-1247*

Why is there a saturation pattern of fitness?

1. Accumulation of negative mutations
2. Negative combination of beneficial mutations

Removing negative mutations

Strains	Genotype
5DU	CEN.PK113-5D; pSP-GM2(*URA3*)
RAU	CEN.PK113-5D ; pSP-GM2(*URA3*); ***RAS2*** **Lys 77**
RBU	CEN.PK113-5D ; pSP-GM2(*URA3*); ***RAS2*** **Tyr112**
EBU	CEN.PK113-5D ; pSP-GM2(*URA3*); ***ERG5*** **Pro 370**

Combination of beneficial genetic change
Site-directed mutations + *PGM2* overexpression

Strains	Genotype
PGM2	CEN.PK113-5D ; pPGM2(***PGM2***)
RA-P	CEN.PK113-5D ; pPGM2(***PGM2***); ***RAS2*** **Lys 77**
RB-P	CEN.PK113-5D ; pPGM2(***PGM2***); ***RAS2*** **Tyr112**
EB-P	CEN.PK113-5D ; pPGM2(***PGM2***); ***ERG5*** **Pro 370**

Appl. Environ. Microbiol. (2012) **78**:7579-7586

Phenotypic improvement

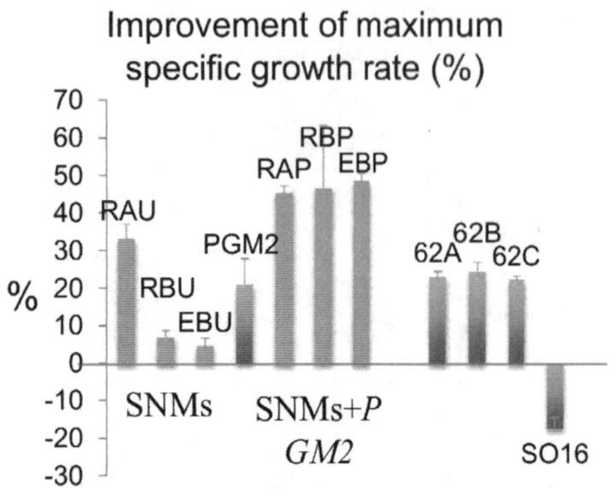

- Single nucleotide mutations (SNMs) results in improved galactose uptake
- Combination of SNMs with over-expression of *PGM2* result in the same (or larger) increase in the specific growth rate on galactose as in the adaptively evolved mutants

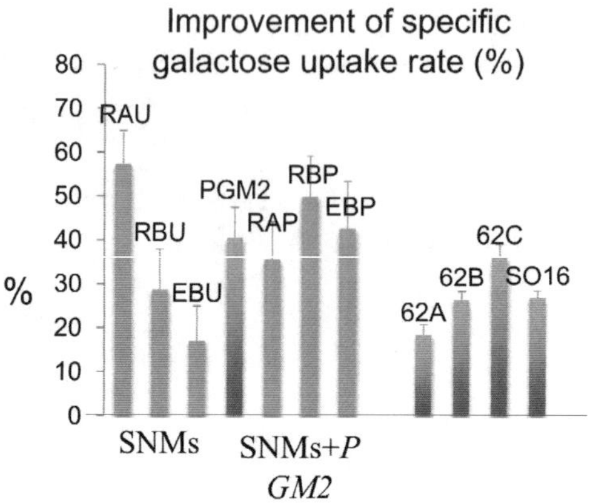

Appl. Environ. Microbiol. (2012) **78**:7579-7586

37

Molecular basis of the mutations

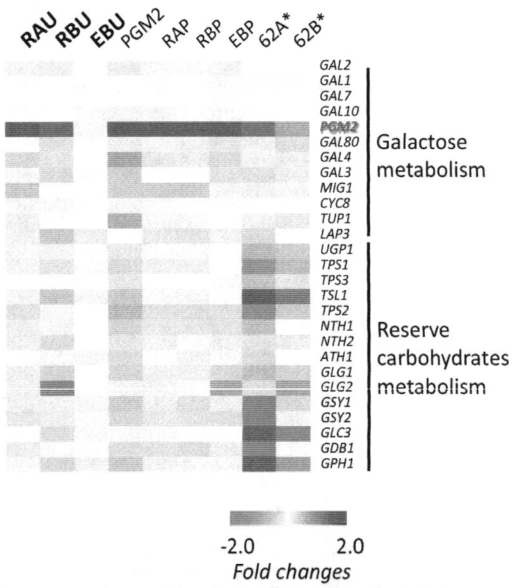

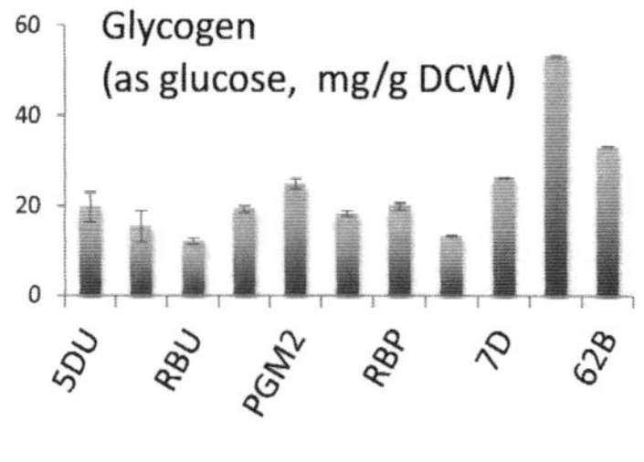

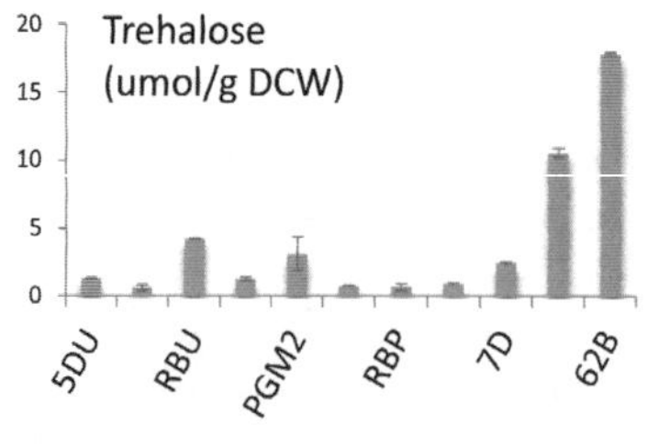

Appl. Environ. Microbiol. (2012) **78**:7579-7586

38

Trade-Off in Evolution

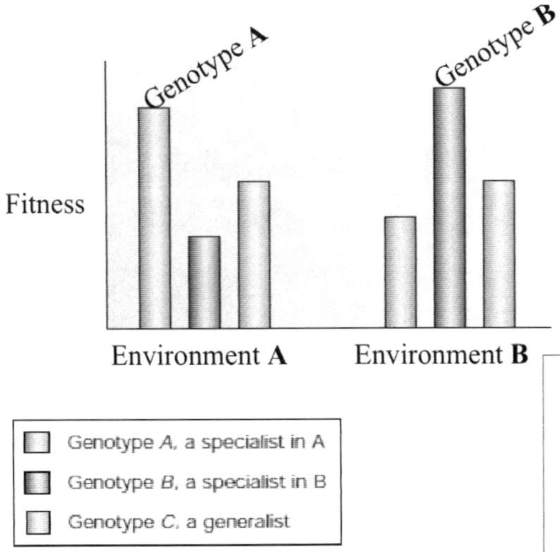

Elena, S.F. & Lenski, R.E. 2003 Nat. Rev. Genet. 4:457–469

Adapative evolution may result in trade-off in terms of other phenotypic traits

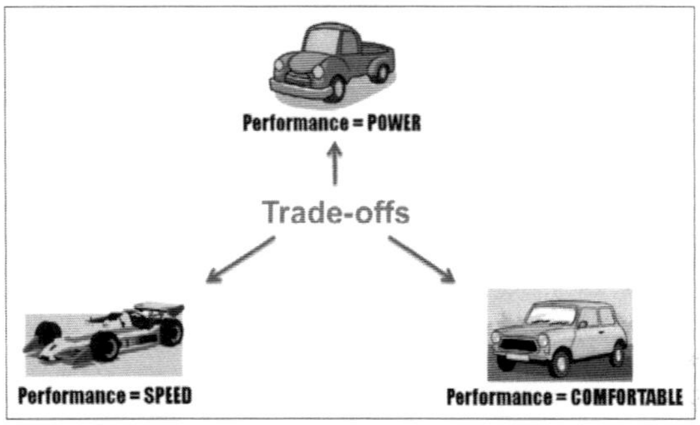

39

Mechanisms of Evolutionary Trade-Offs: AP vs. MA

Antagonistic Pleiotropy (AP)

The same mutation that is beneficial in one environment is harmful in the other.

Mutation Accumulation (MA)

Different mutations produce the direct fitness gain and the correlated losses.

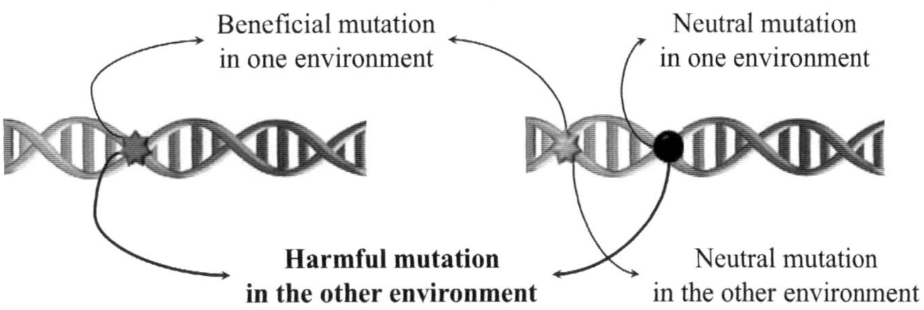

Identification of mechanism type is important for Metabolic Engineering

40

Trade-off in Galactose and Glucose Utilization

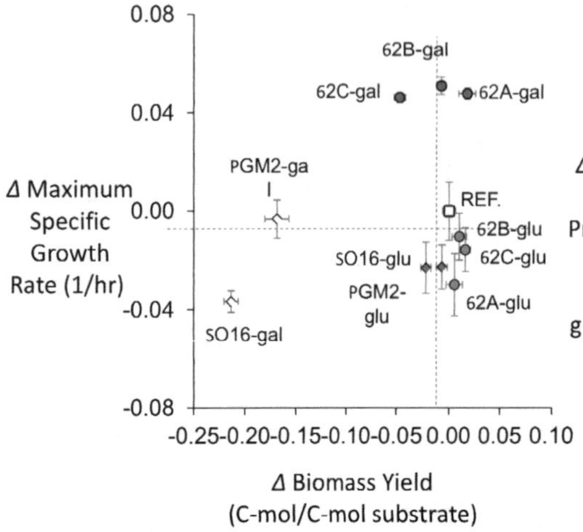

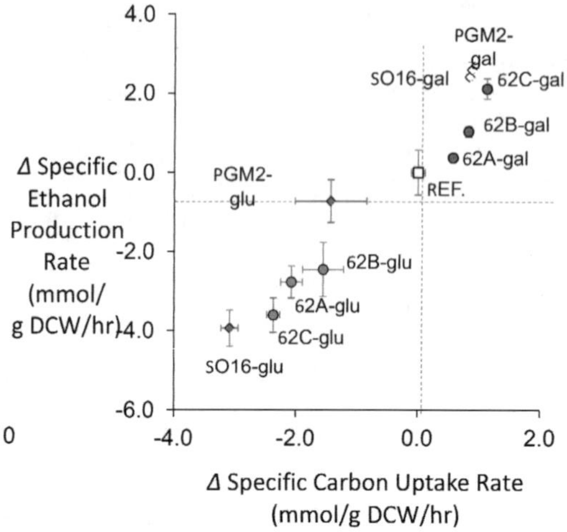

Submitted for publication 41

Transcriptome Analysis

The largest changes in the transcriptome is associated with the change in carbon source

Similar clustering among the mutants grown on the two carbon sources

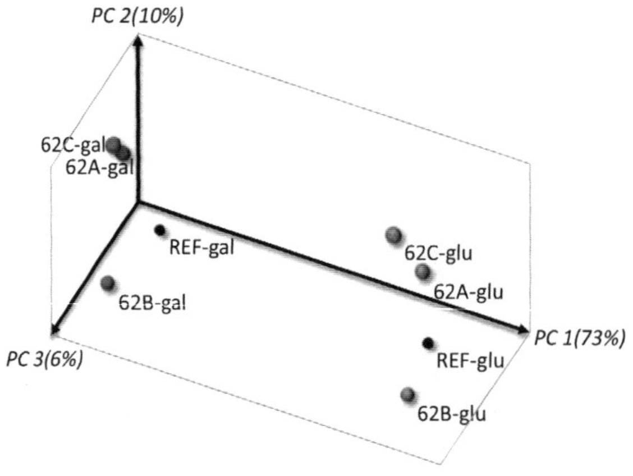

Submitted for publication 42

Transcriptome Analysis

sys²bio www.sysbio.se

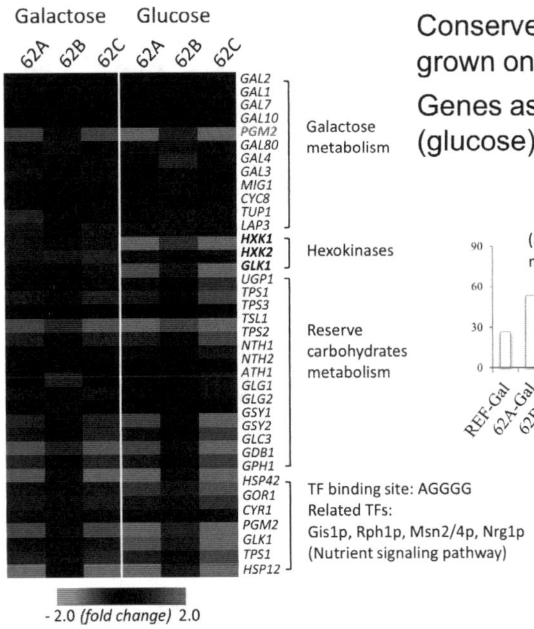

Conserved changes among the mutants grown on the two carbon sources

Genes associated with carbon catabolite (glucose) repression

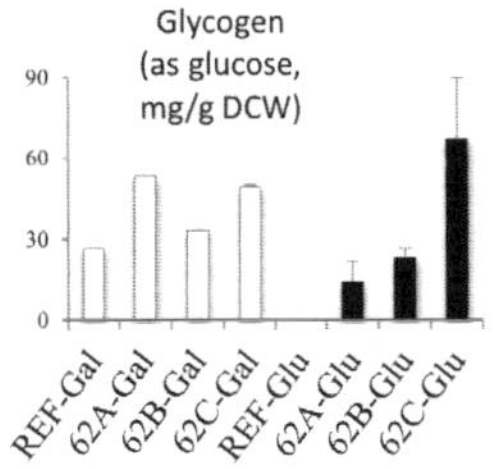

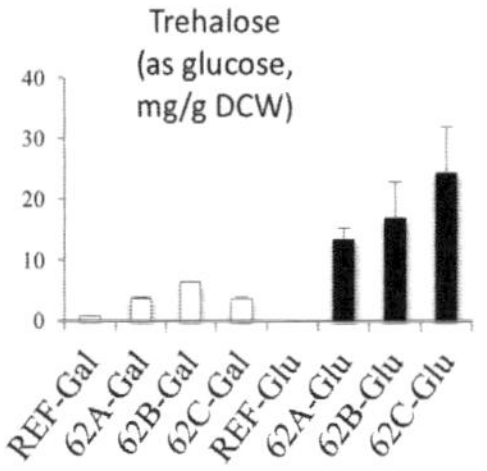

Submitted for publication

CHALMERS 43

Transcriptome Analysis

sys²bio www.sysbio.se

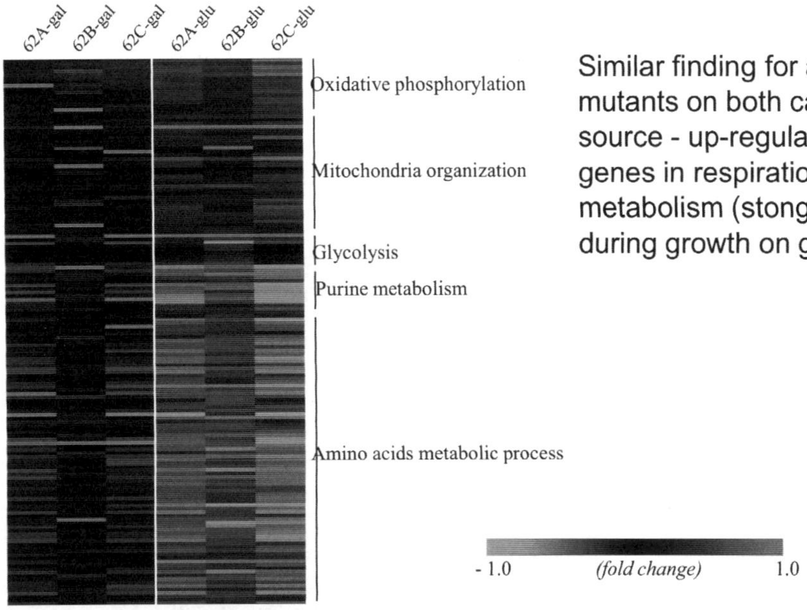

Similar finding for all three mutants on both carbon source - up-regulation of genes in respiration metabolism (stonger effect during growth on glucose)

Submitted for publication

CHALMERS 44

Trade-off in Galactose and Glucose Utilization

sys²bio www.sysbio.se

We also observed trade-off for the strains with single nucleotide mutations

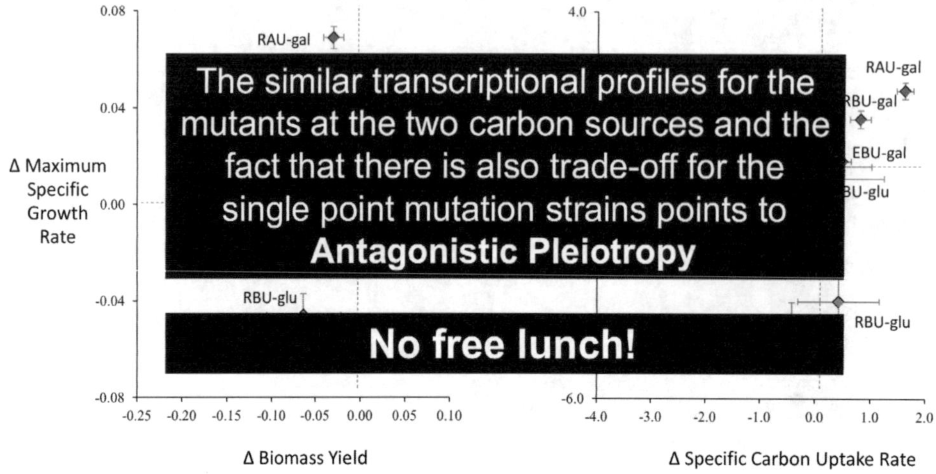

Submitted for publication 45

CHALMERS

Mechanism

sys²bio www.sysbio.se

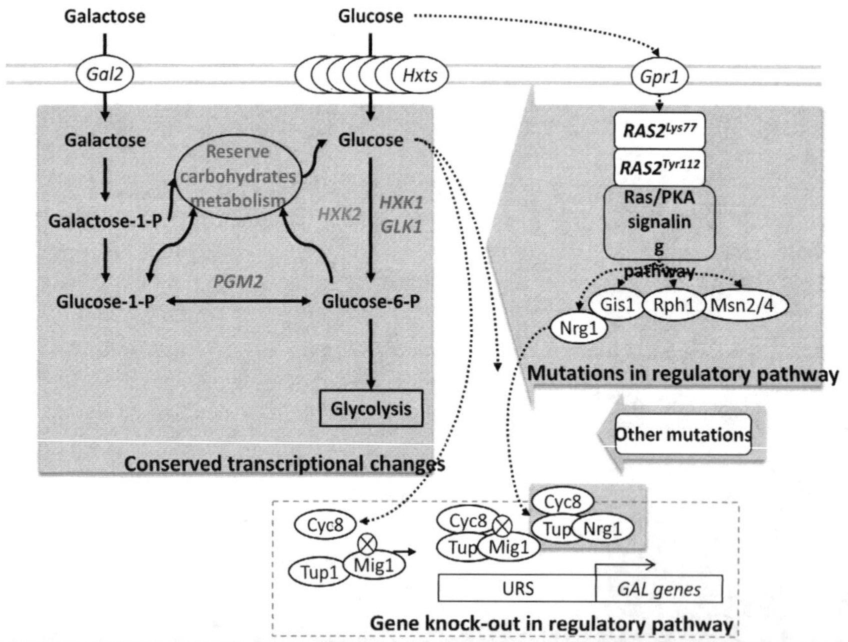

Submitted for publication 46

CHALMERS

Conclusions

Characterization of adaptively evolved mutants by integrative systems biology may expands our understanding of regulation that can be used for advancing metabolic engineering

Three lessons obtained from the study of galactose metabolism relevant for metabolic engineering:

1. Non-canonical genetic change may appear
   → possibility to find new ME targets
2. Accumulation of non-necessary changes
   → inverse metabolic engineering can be used to insert only driving mutations
3. Trade-off in traits
   → classification of trade-off mechanism (Antagonistic Pleiotropy or Mutation Accumulation) enables evaluation of whether the trade-off can be avoided through inverse metabolic engineering

47

Acknowledgement

Modeling work:
Tobias Österlund
Rasmus Ågreen

Succinate story:
Jose Manuel Otero
Kiran Patil
Donatella Cimini

Galactose story:
Simon Ostergaard
Christoffer Bro
Kuk-Ki Hong
Wanwipa Vongsangnak

www.sysbio.se

48

Engineering an alternate carbon fixation pathway

Matt Mattozzi
Wyss Institute for Biologically Inspired Engineering
Pam Silver Lab, Harvard Medical School

International Conference on Biomolecular Engineering
Fort Lauderdale
Tuesday, 15 January 2013

WYSS INSTITUTE

Goals

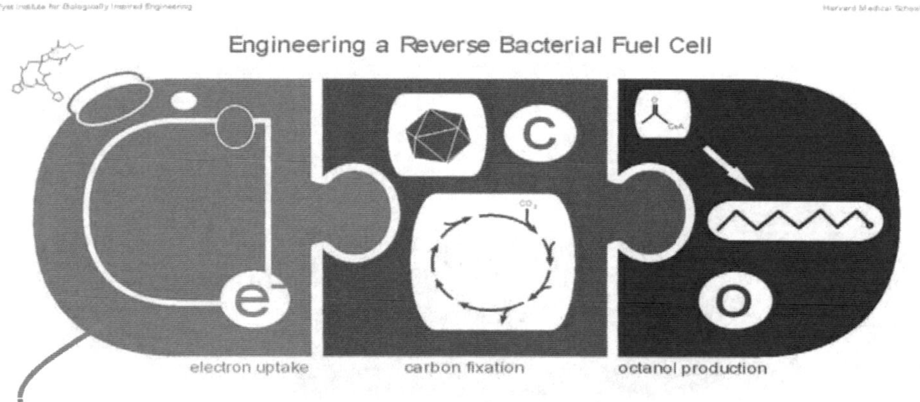

- Express unique carbon fixation cycle from *Chloroflexus aurantiacus* in *E. coli*
  - Turn a heterotroph into an autotroph?
- Integrate with biofuel production capacities

Agapakis/Boyle

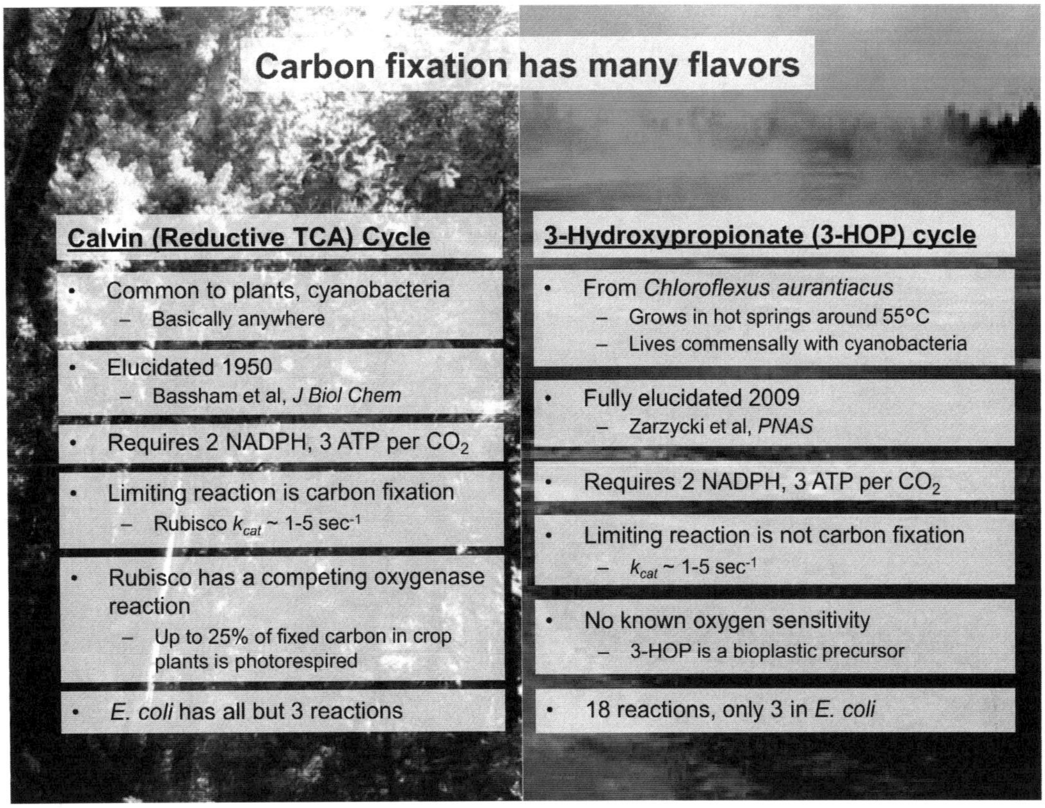

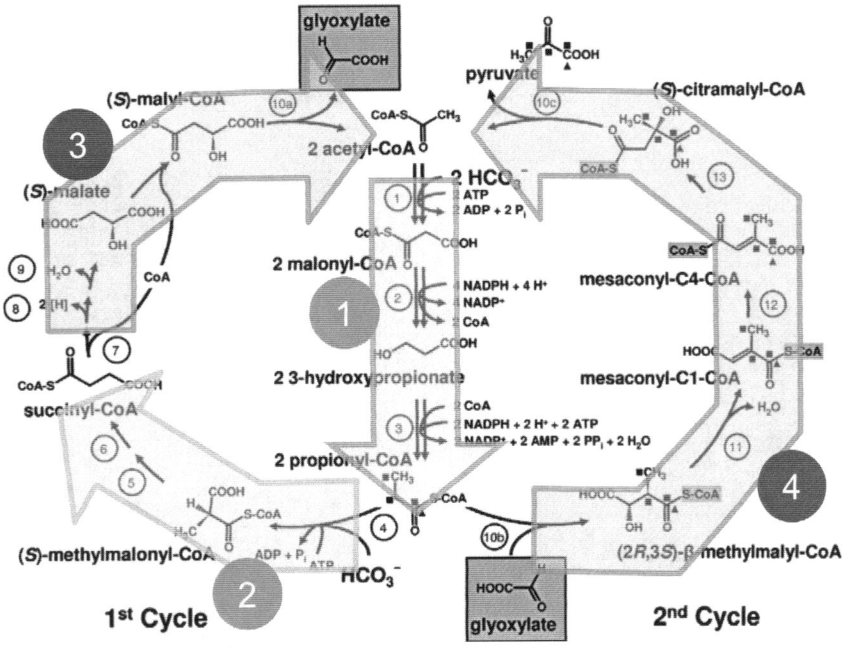

- Bi-cycle divided into 4 sub-pathways for functional genetic tests

Zarzycki et al 2009

A biosensor for propionyl-CoA production

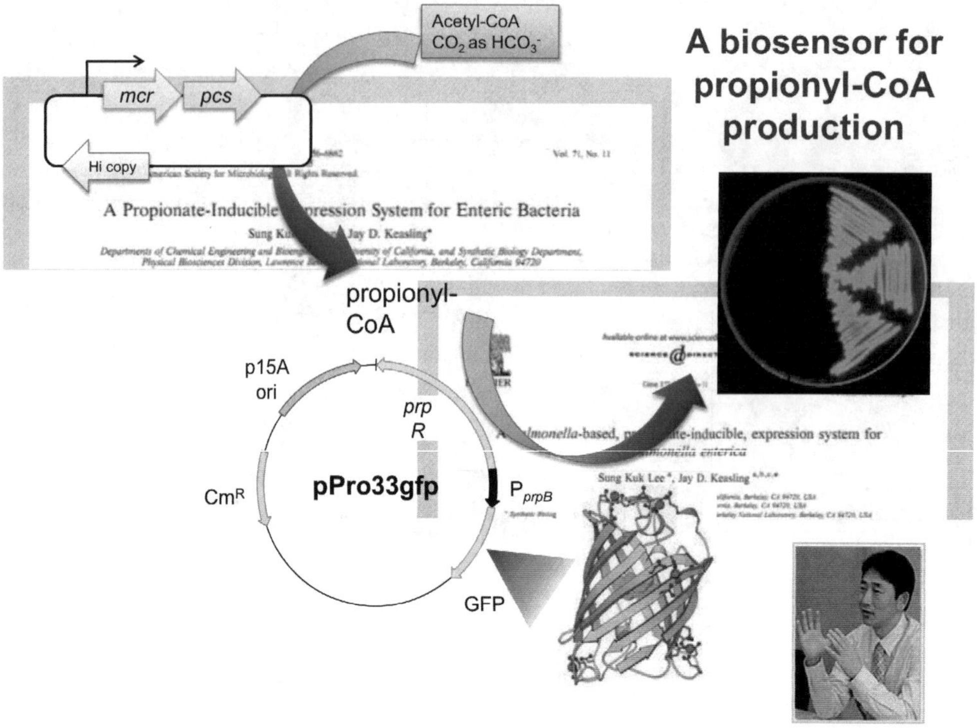

Middle pathway makes a lot of propionate

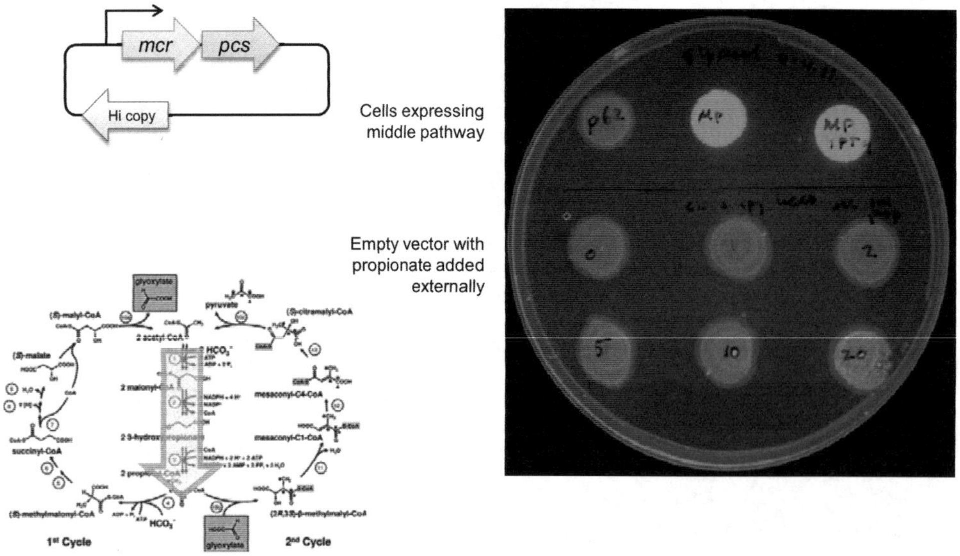

Propionate is potent bacterial inhibitor

- Exact method not well known
  - Likely inhibits pyruvate dehydrogenase

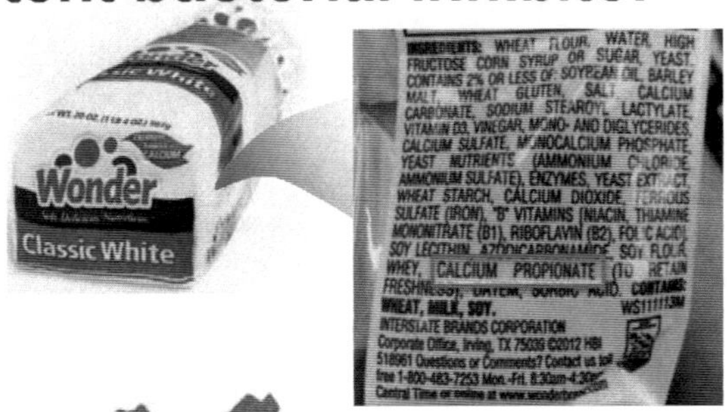

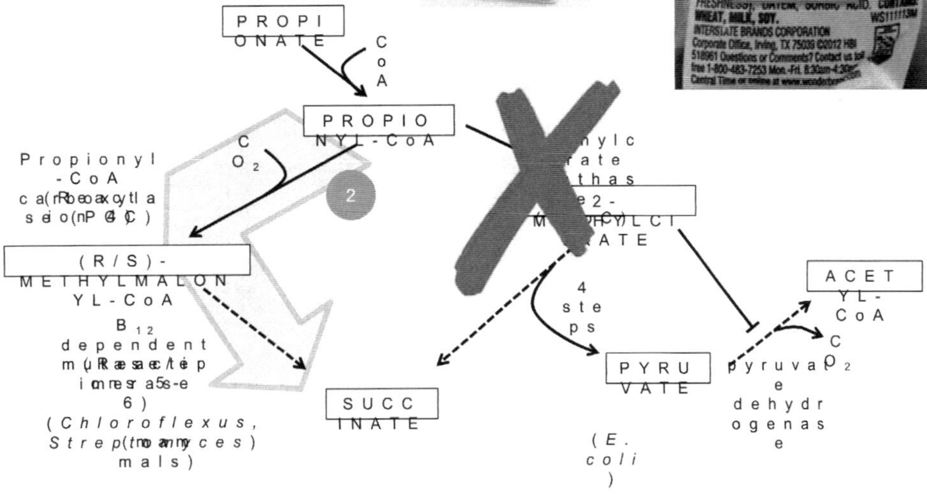

Two assays

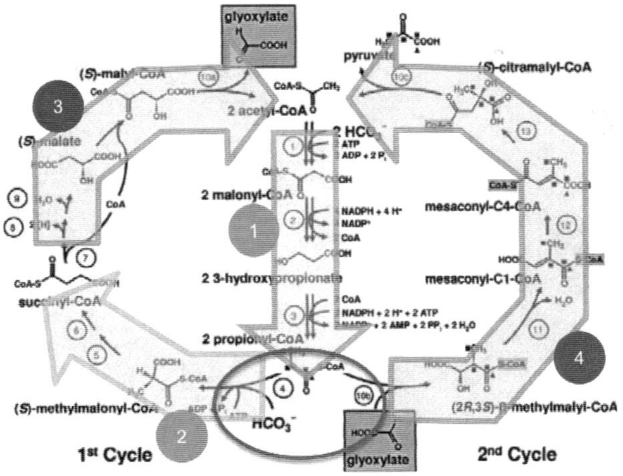

- **Propionate detoxification**
  - Relieves poisoning by propionate
  - Less glycerol/glucose: better expression of pathway

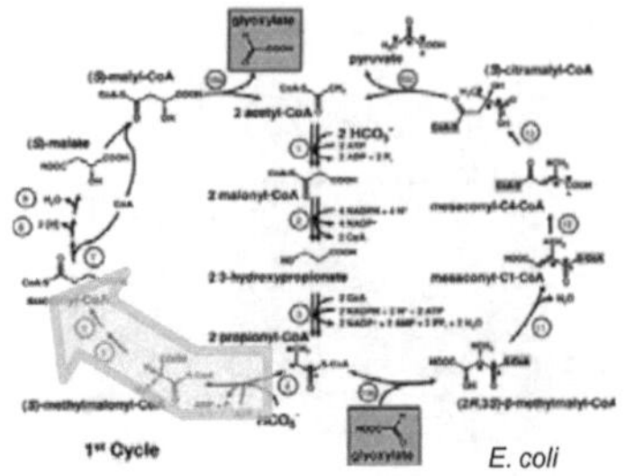

Streptomyces PCC detoxifies the fastest

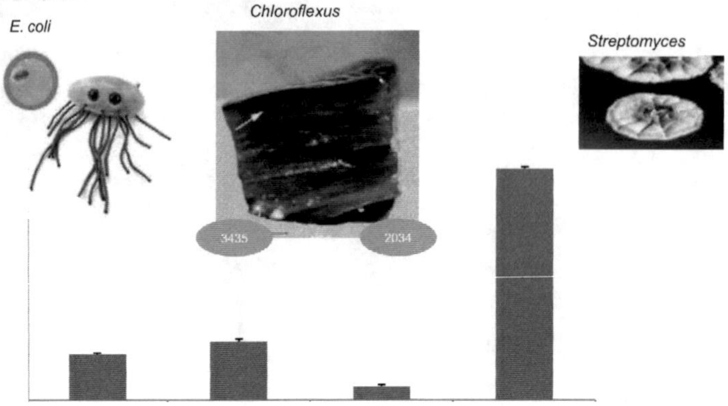

BW25113 cells after 24 h
M9 0.14% glucose, 0.26% propionate.

Right side detoxifies propionate

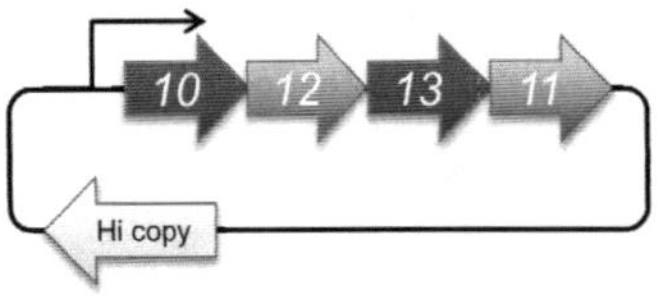

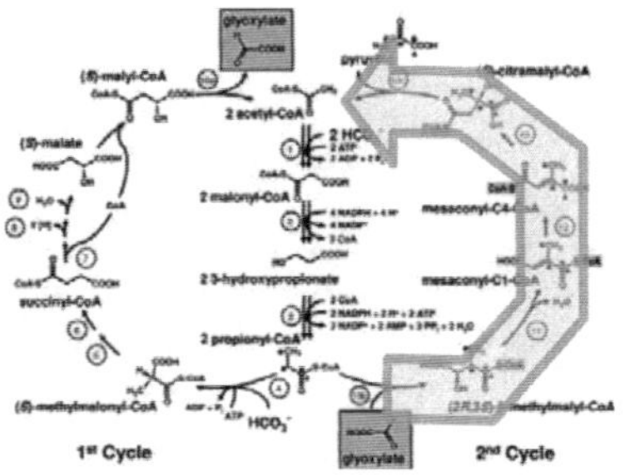

- Slower enzymes in synthetic operon spaced closer to the promoter wi[redacted]

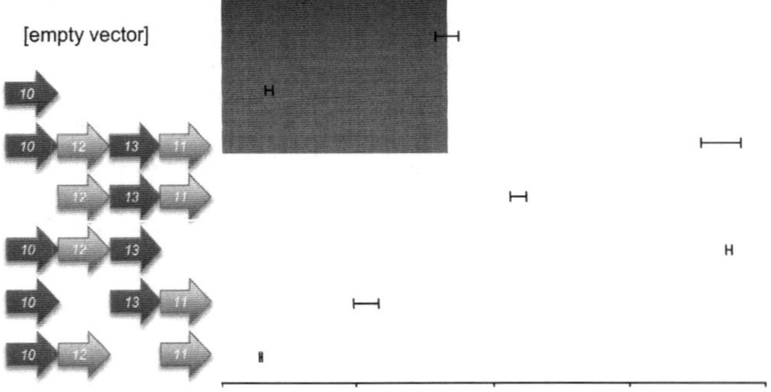

  - *prpC-* cells grown in M9 0.32% propionate 0.08% glucose, 48 h

Host *E. coli* context

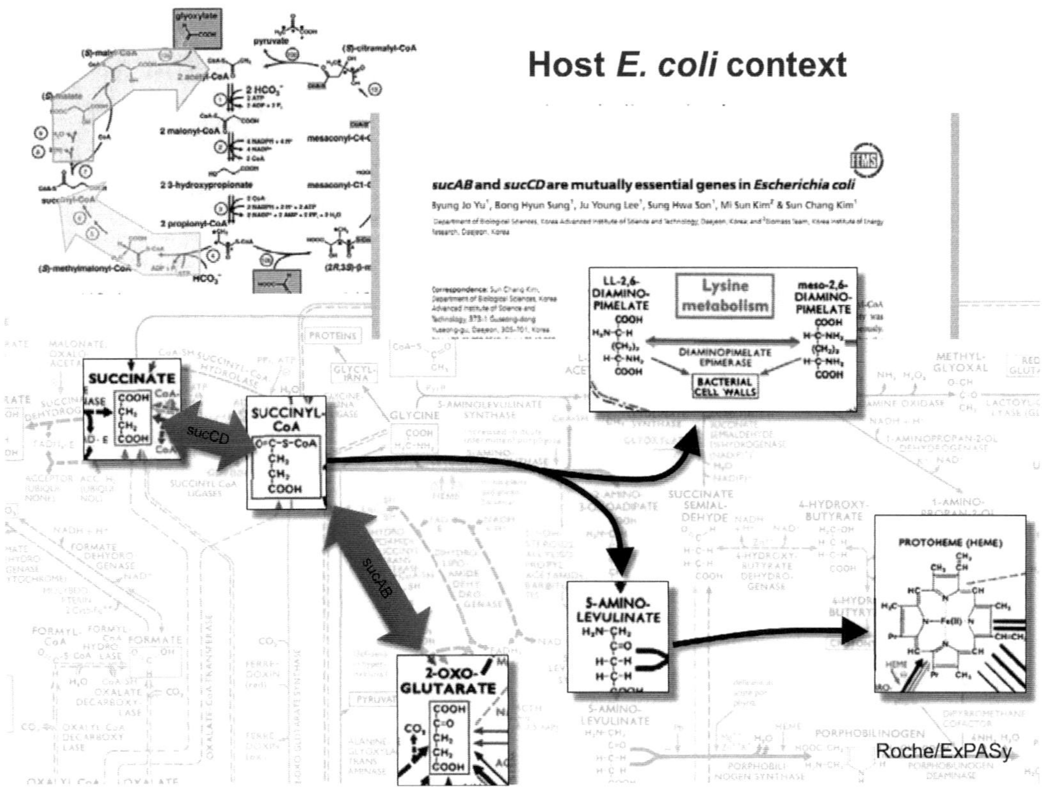

Two assays

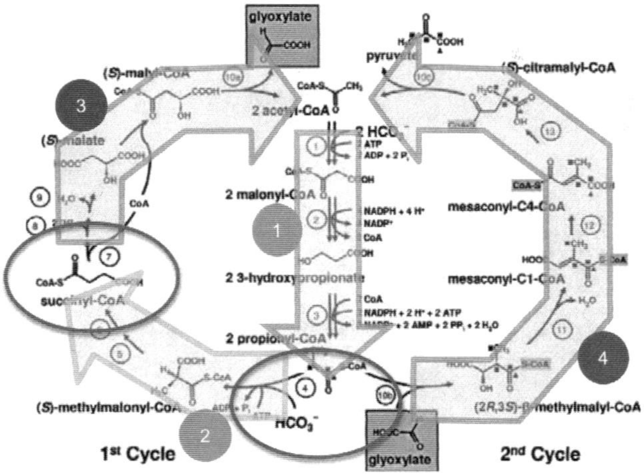

- **DAP titration**
  - *sucAD*[-] cells require diaminopimelic acid (DAP) for cell walls
  - Less DAP: better expression of pathway

- **Propionate detoxification**
  - Relieves poisoning by propionate
  - Less glycerol/glucose: better expression of pathway

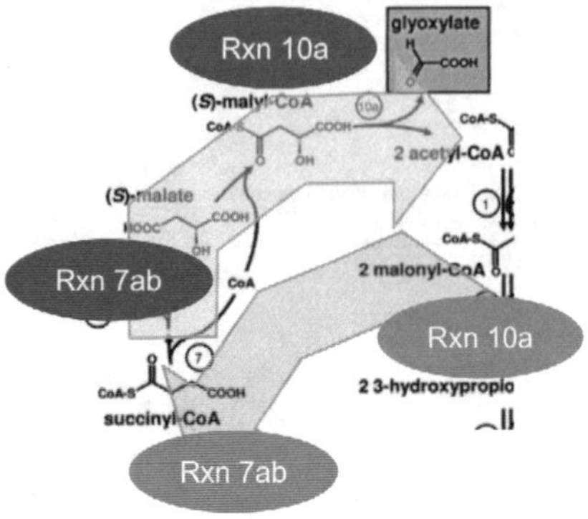

Pathway 3 complements *sucAD*⁻'s DAP requirement

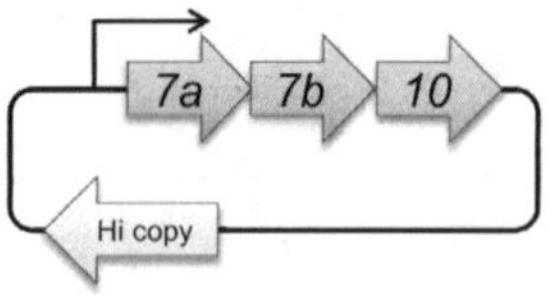

- Hope that the reaction also runs in reverse to complement a succinyl-CoA auxotroph

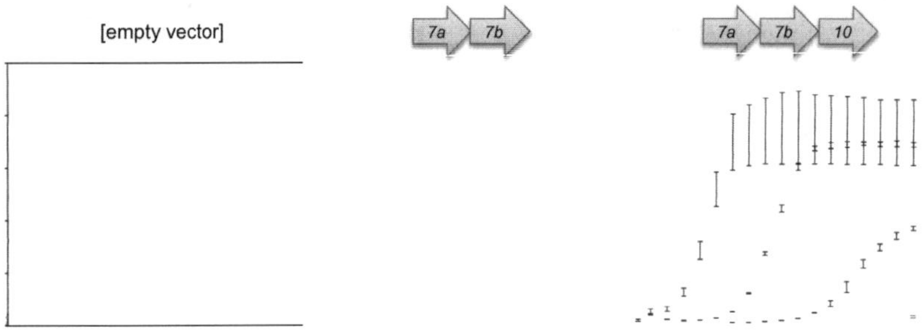

Expressing multiple pathways does not result in autotrophy

	WT	Sub-pathways on high-copy plasmids								
Carbon source	No plasmid	Empty vectors	1	1,2	4	1,4	1,2,4	3,4	1,3,4	1,2,3,4
bicarbonate only (autotrophic)	-	-	-	-	-	-	-	-	-	-
+ acetate (+ control)	++	++	++	++	++	++	++	++	++	++
+ propionate (toxic!)	-	-	-	+	+	+	+	++	+	++

	WT	Sub-pathways chromosomally integrated								
Carbon source	Not trans-formed	Empty vectors	1	1,2	4	1,4	1,2,4	3,4	1,3,4	1,2,3,4
bicarbonate only (autotrophic)	-	-	-	-	-	-	-	-	-	-
+ acetate (+ control)	++	++	++	++	++	++	++	++	++	++
+ propionate (toxic!)	-	-	-	-	-	-	-	++	+	+

- 2 weeks growth on M9 noble agar plates

Where we're at

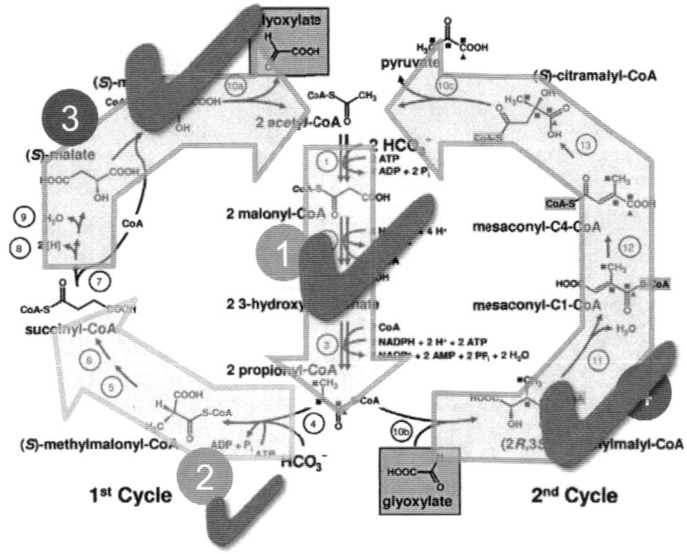

- Combined, no autotrophy
- We can detoxify propionate
- Need to optimize reducing equivalents
- Integrate with a bioproduct synthesis

Thank you!

Jeff Way **Pam Silver** **Mathias Voges** **Marika Ziesack**

Wade Hicks
Colby Stoddard

Avi Robinson-Mosher
Lev Shaket

Tom Ferrante

Blaine Pfieffer, U Buffalo

Anna Chen
Drew MacKellar
Joe Torella
Tyler Ford
Danny Ducat
Buz Barstow

Patrick Boyle
Devin Burrill
Mara Inniss

Jen Sheen, MGH
Jen Bush, MGH

WYSS INSTITUTE

arpa·e

Engineering Anaerobic Gut Fungi for Lignocellulose Breakdown

Michelle A. O'Malley
Assistant Professor
Department of Chemical Engineering
University of California, Santa Barbara

Recalcitrant Biomass Hinders Production of Valuable Products from Cellulose

UCSB

Cellulosic biomass is an attractive renewable resource for biofuels, bioplastics, etc.

3 enzymatic components required:

- Endoglucanases
- Exoglucanases
- β-glucosidases

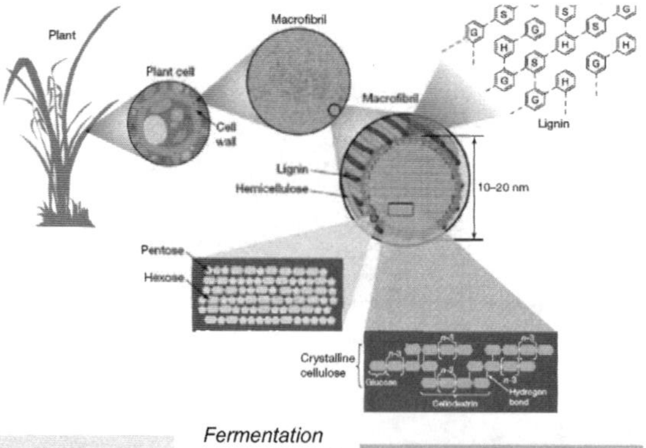

Pretreatment, Enzymatic Hydrolysis

Cellulose-rich Feedstock → **Sugar** → *Fermentation* → **Value-added Product**

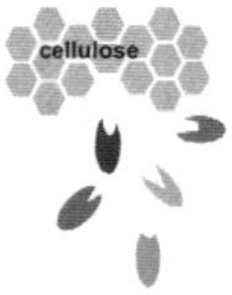

Purified cellulases (i.e. from *T. reesi, A. niger*)

Major Roadblocks:

- Recalcitrant nature of crude biomass (presence of lignin and other inhibitors)
- Costs associated with enzyme production and catalytic performance

2

Rubin, E. M. Nature 454 (2008) 841- 845.

What can we learn from nature?

- Several microbiomes have evolved enzymes over millions of years to degrade plant material
- Many attractive enzymes remain undiscovered
- Cellulolytic anaerobic fungi are attractive targets
  - Substrate breakdown is both chemical *and* mechanical
  - Digests crude, lignin-rich substrates
  - Enzymes are likely compatible with an industrially-relevant host (yeast – *S. cerevisiae*)

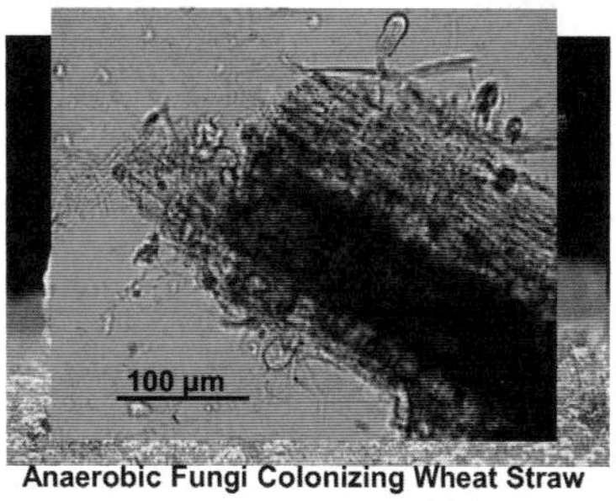

Anaerobic Fungi Colonizing Wheat Straw

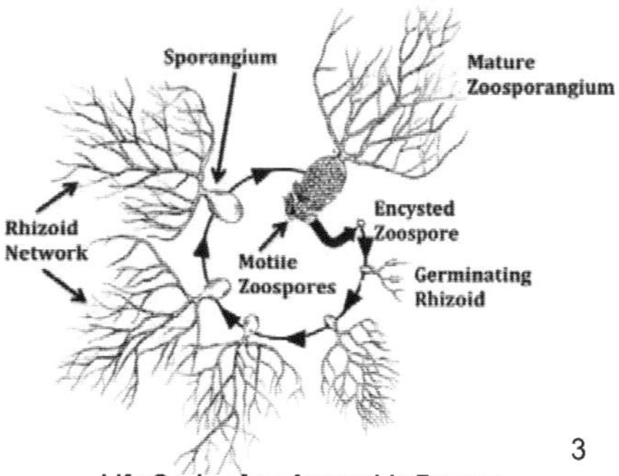

Life Cycle of an Anaerobic Fungus

R. Doi, Ann. N.Y. Acad. Sci. 1125:267-279 (2008).
O'Malley MA, Theodorou MK, Kaiser CA. *Environmental Progress*, 31(1): 37-46 (2012).

3

Anaerobic Fungi are Powerful Cellulolytic Microbes

- Fungi can easily depolymerize crystalline cellulose (Whatman #1 Filter Paper)
- Similar results are observed for cellulosic substrates with high lignin content

Filter Paper Alone

3-4 Days Post-inoculation

7-8 Days Post-inoculation

Why are these fungi so good at breaking down cellulose?

4

O'Malley MA, Zhong Y,Theodorou MK, Kaiser CA. *In preparation.*

Fungal Cellulosomes Facilitate Targeted, Synergistic Cellulose Hydrolysis

UCSB

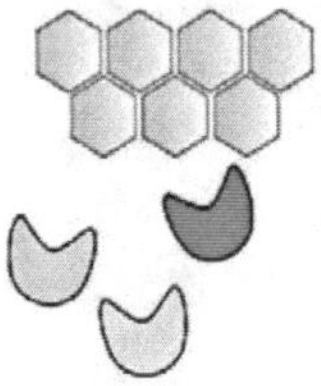

Aerobes Secrete Free Enzymes

➢ Enzymatic modules are fused into a protein scaffold via **dockerin-cohesin** interaction to form > MDa complexes

➢ Recent genomic sequencing of *Piromyces sp E2* by the DOE-JGI has revealed 360+ dockerin-tagged enzymes

5

Flint HJ, Bayer EA, et al. Nature Reviews Microbiology, 6 (2008) 121-131.

Heterologous Expression Doesn't (Completely) Solve the Problem

UCSB

✓ Gut fungal enzymes can be made in a simpler yeast (*S. cerevisiae*)
✓ Enzymes are post-translationally modified
✓ Some proteins are secreted, though there are secretion bottlenecks
✓ Activity of recombinant enzymes is <u>variable</u>

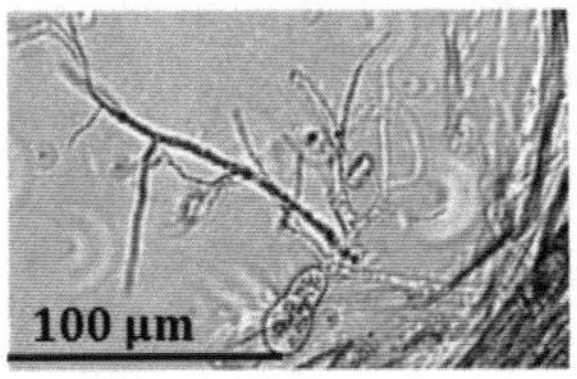

Gut Fungal Cellulases and Cellulosomes

?

→

Yeast

We need to know more…
➢how are gut fungal enzymes made, and how are they regulated?
➢how is structure connected to cellulosome function?
➢in what ways can these organisms be engineered?

6

O'Malley MA, Theodorou MK, Kaiser CA. *Environmental Progress*, 31(1): 37-46 (2012).

Objectives

Overall Goal: Accelerate the development of renewable lignocellulosic biofuels and value-added products

Approach: Learn from nature to engineer cellulases and lignocellulolytic gut fungi with novel functions

- Isolate and screen gut fungi from native microbiomes for attractive lignocellulolytic properties
- Construct metabolic models to describe mechanisms responsible for enzyme regulation
- Develop genetic engineering strategies to manipulate gut fungi

7

What can we learn by studying gut fungi *directly*?

- Cultivation of anaerobic fungi is inherently difficult due to strict medium requirements and short lifetimes
  - Growth temp ~39 - 45°C
  - Life cycle 3 - 4 days
  - Frequent subculture required
- 6 known genera of anaerobic fungi
  - *Neocallimastix*, *Piromyces*, *Caecomyces*, *Orpinomyces*, *Cyllamyces*, and *Anaeromyces*
- An exceptionally small research community works with these fungi
  - Trained by Michael K. Theodorou (Aberystwyth/Durham University, UK)

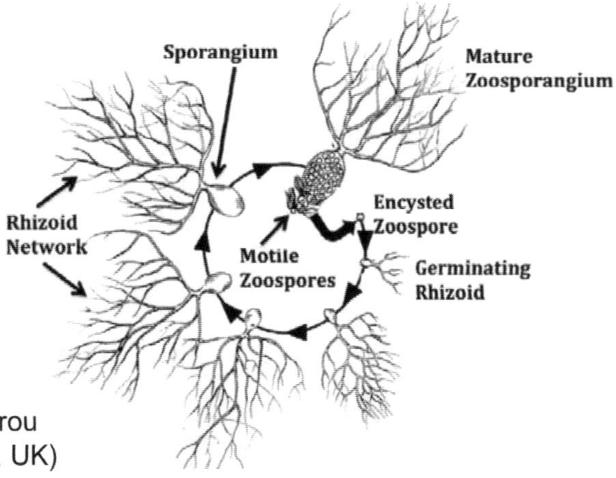

Life Cycle of an Anaerobic Fungus

Orpin, et al. Anaerobic Fungi, 2005.
Trinci, et al. 1994.
R. Doi, Ann. N.Y. Acad. Sci. 1125:267-279 (2008).

8

Anaerobic Gut Fungi are Isolated from Crude Source Material

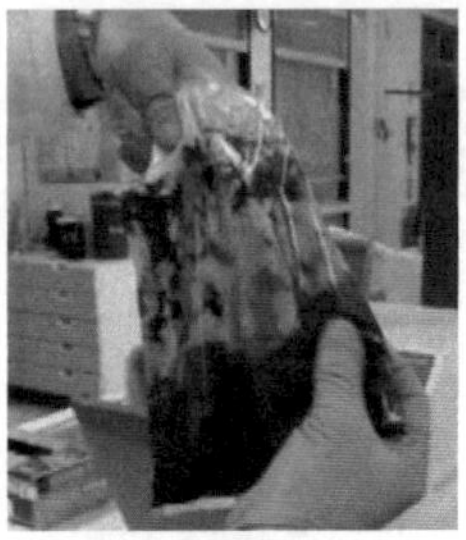

Animal Digesta

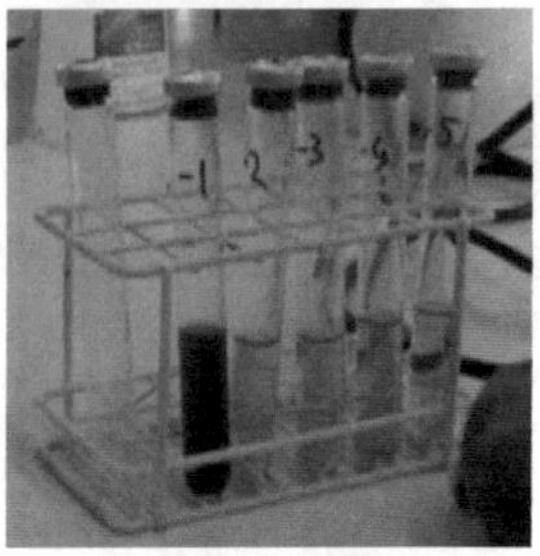

Anaerobic Medium

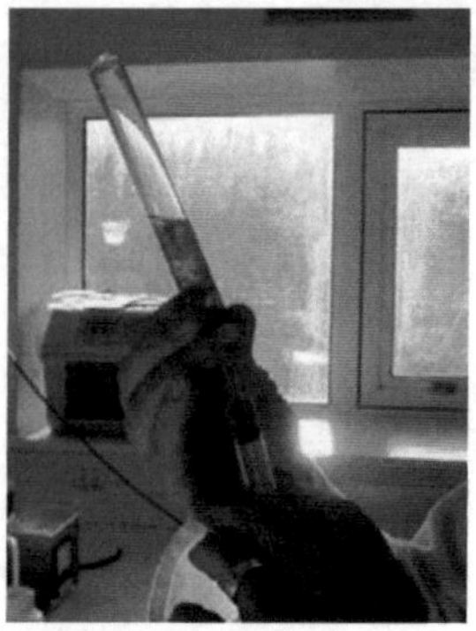

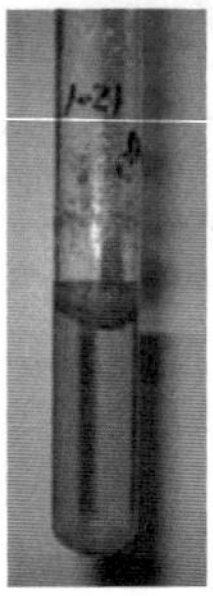

Viable Fungi

Culture conditions are remarkably complex for these microbes

- Requires rumen fluid, redox indicators, strict anaerobic environment, growth between 39º-42ºC
- Fungi produce fermentation gases, leading to floating substrate
- Organisms are incredibly difficult to cryo-preserve, and are generally sustained through repeated sub-culture

9

A Novel Gut Fungal Isolate is Obtained from a Horse

- Fungi were extracted from a horse named "Finn" from Verrill Farm Stables (Concord, MA)
- "Finn fungi" were supported on reed canary grass supplied by the USDA
    - Full growth cycle is seen in 5 – 6 days
    - Frequent sub-culture is required; active growth is revealed via fermentation gas generation (and floating substrate plug)

Finn
(courtesy of Verrill Farm)

Inoculation | 2 – 4 days | 5 – 6 days | > 6 days

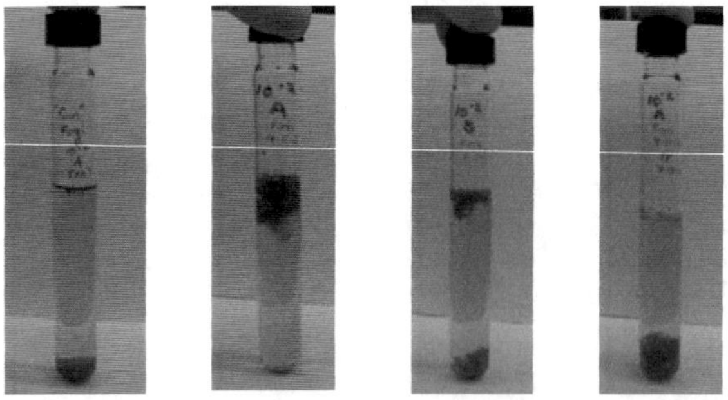

10

Fungal Growth Depends on the Available Carbon Substrate

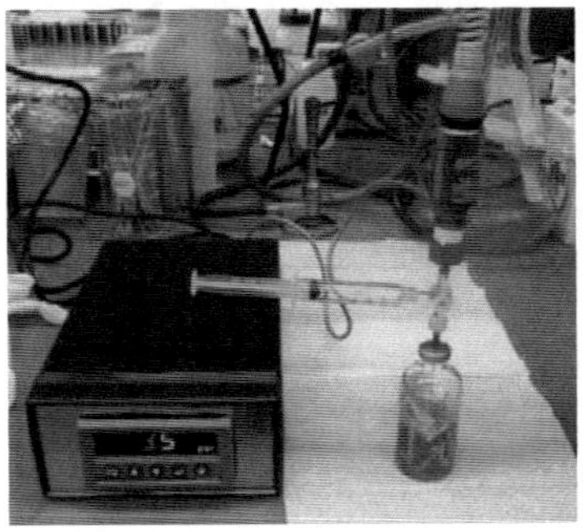

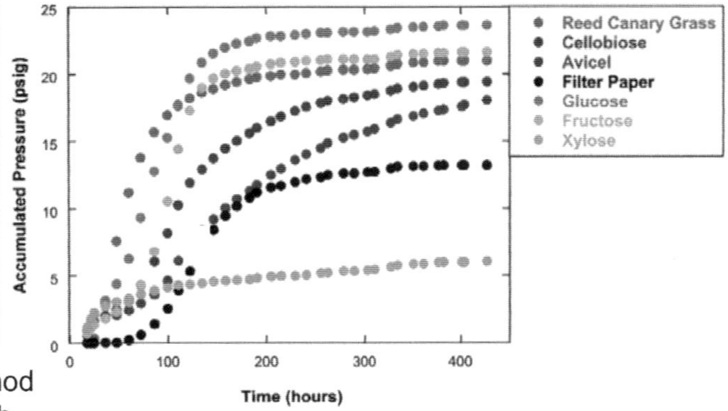

➢ A pressure transducer method can be used to quantify growth

➢ Finn fungi readily proliferate on cellulose-rich substrates (reed canary grass, Avicel, filter paper), cellobiose, glucose, and fructose

➢ 5-carbon sugars (xylose, arabinose), and several 6-carbon sugars do not sustain continued growth

11

Theodorou MK, Davies DR, et al. Microbiology, 141 (1995) 671-678.
O'Malley MA, Theodorou MK, Kaiser CA. *Environmental Progress*, 31(1): 37-46 (2012).

Finn Fungus is from the Genus *Piromyces*

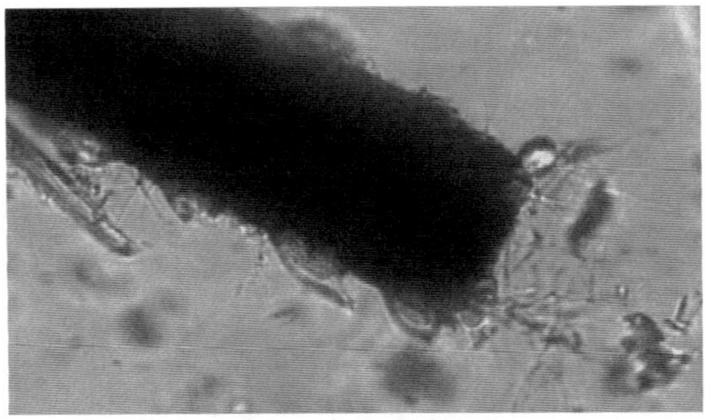

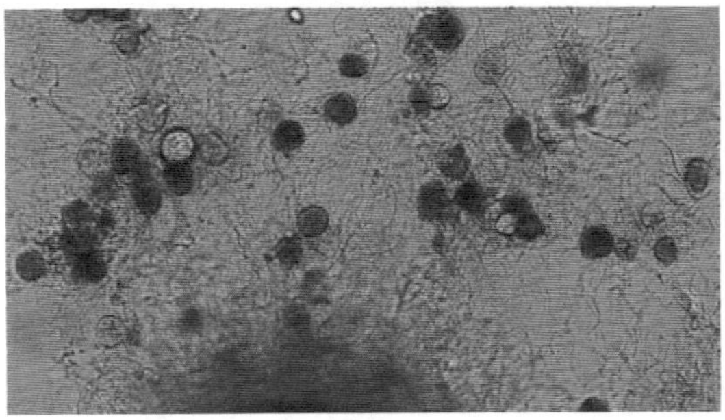

➢ ITS fingerprinting and morphology analysis indicate that the Finn fungus is a unique isolate related to *Piromyces*

➢ Isolate named *Piromyces sp finn*

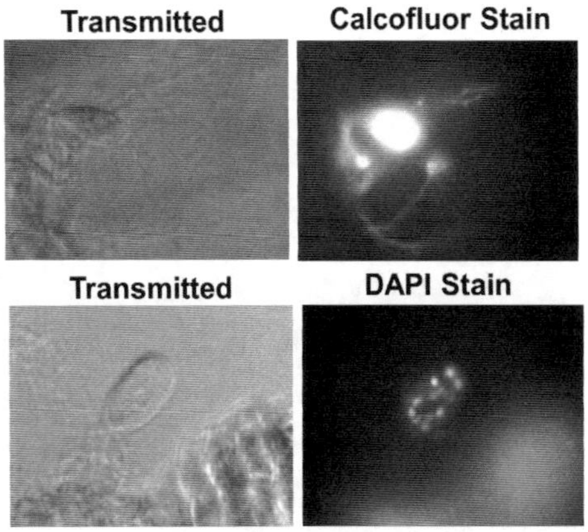

Enzyme Discovery: Why Sequencing a Genome is Not Sufficient

UCSB

DNA —*transcription*→ (mRNA) —*translation*→ **Protein**

To date, almost all efforts aimed at novel cellulase discovery have focused on genomic DNA sequencing

Pros:

✓ Identification is easily carried out through homology queries

✓ Copy number can be determined

✓ Genetic regulatory regions are easily located

Cons:

x DNA ≠ mRNA due to alternative splicing in fungi

x Radically divergent proteins are missed

x Protein production levels are unknown

In order to identify/characterize cellulolytic proteins, it is critical to examine the mRNA from a growing organism (transcriptome).

Next Generation Sequencing Facilitates *de novo* Transcriptome Assembly

UCSB

Analysis of Total RNA from Fungi Grown on Reed Canary Grass

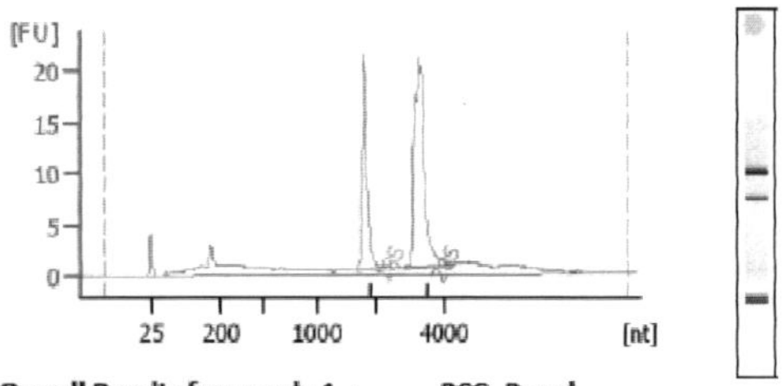

Overall Results for sample 1 : RCG Broad

RNA Area: 135.4
RNA Concentration: 157 ng/µl
rRNA Ratio [28s / 18s]: 1.7
RNA Integrity Number (RIN): 9.1 (B.02.07)
Result Flagging Color:
Result Flagging Label: RIN: 9.10

- High-quality RNA can be extracted
- First anaerobic fungal transcriptome assembled using Trinity platform (Regev Group, Broad Institute)
  - 150 million paired-end reads (Illumina HiSeq)
  - Strand-specificity of the library facilitates assembly

- Over 27,000 novel transcripts were assembled *de novo,* without the need for a reference genome!

14

Grabherr MG, Haas BJ, et al. Nature Biotechnology 29 (7) 644 – 652, (2011)

Bioinformatic Workflow: Finding Cellulases and Other Enzymes

1. Extract ORFs from the transcripts
2. Translate the ORF to protein sequence
3. Align transcriptome against known sequences
   - ➢ Fungal dockerins
   - ➢ Specific GH families
4. Validate hits w/ web servers
   a) NCBI Conserved Domain Database
   b) EMBL-EBI InterPro
   c) BLAST
5. Analyze and compare hits

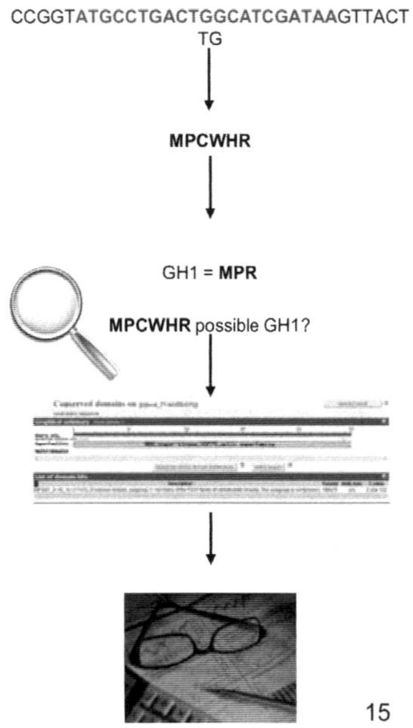

15

Kevin Solomon

Novel Fungal Cellulases are Found within the Transcriptome

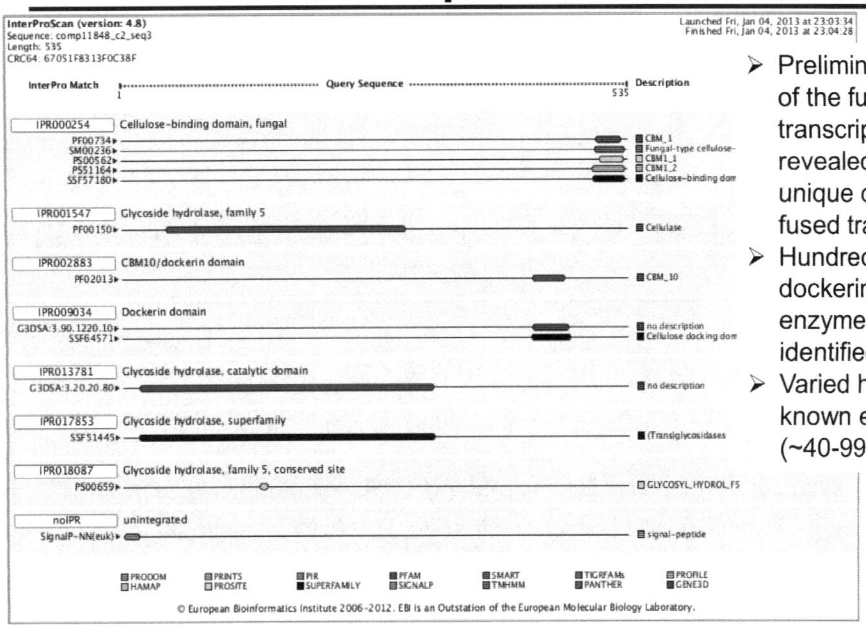

- ➢ Preliminary analysis of the fungal transcriptome has revealed 300+ unique dockerin-fused transcripts
- ➢ Hundreds of non-dockerin tagged enzymes also identified
- ➢ Varied homology to known enzymes (~40-99%)

16

Kevin Solomon

Piromyces sp finn is Rich in Several GH Enzymes and Dockerin-tagged Proteins

	GH1	GH5	GH6	GH9	GH16	GH45	GH3	GH10	GH11	GH43	Polysacc. Deacetyl.	Total
Aspergillus nidulans	1	8	0	0	10	0	13	3	2	14	3	44
Trichoderma ressei	2	3	2	0	1	1	5	1	5	0	0	20
Piromyces sp finn	**1**	**2**	**3**	**3**	**2**	**5**	**4**	**3**	**2**	**4**	**11**	**40**
Piromyces sp. E2	4	1	3	3	0	1	2	0	0	0	0	14
Neocallimastix patriciarum	0	1	1	0	1	0	0	1	5	0	0	9

- The fungal isolate is particularly rich in polysaccharide deacetylases, esterases, and exoglucanases
- Functionality is unique compared to aerobic fungi, and even among other gut fungal isolates

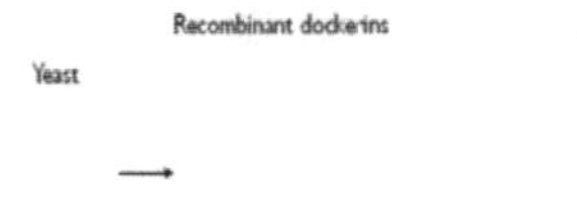

Fungal Dockerin Domains

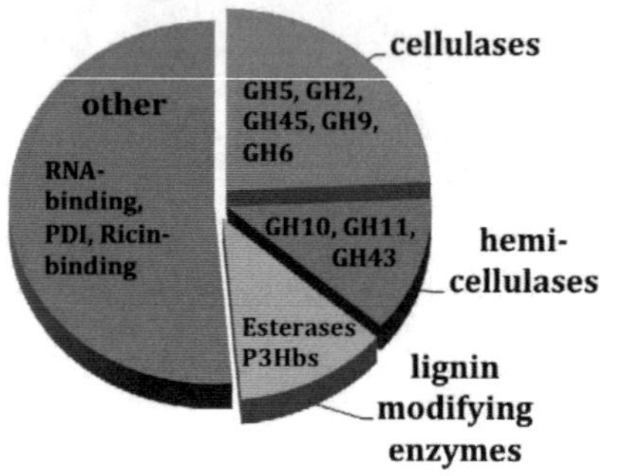

Kevin Solomon

A New Approach for Cellulase Discovery from Anaerobic Gut Fungi

- **Hypothesis:** Cellulosic substrates activate transcription of cellulolytic enzymes and cellulolytic complexes (cellulosomes); monosaccharides repress transcription

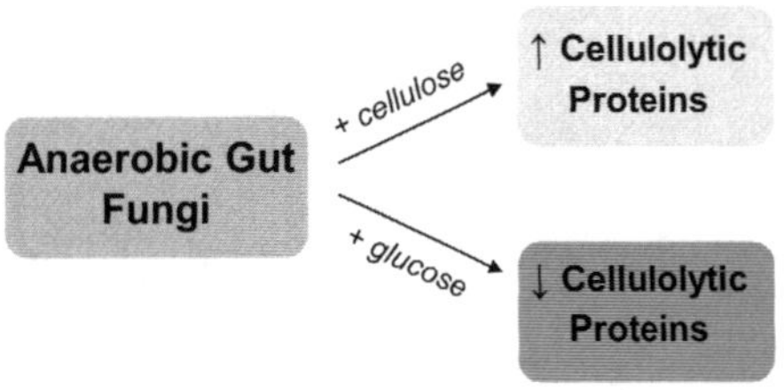

Expected Outcomes:

- Find novel cellulases in gut fungi by deciphering catabolic regulation patterns
- Identify "adapter" domains and/or proteins that are indirectly involved in cellulose hydrolysis
- Potential to identify radically divergent enzymes from known cellulases

18

Aro, N, et al. FEMS Microbiology Reviews 29 (2005) 719- 739.
Dijkerman, et al. Current Microbiology 34 (1997) 91-96.

Catabolic Regulation Occurs in *Piromyces sp finn*

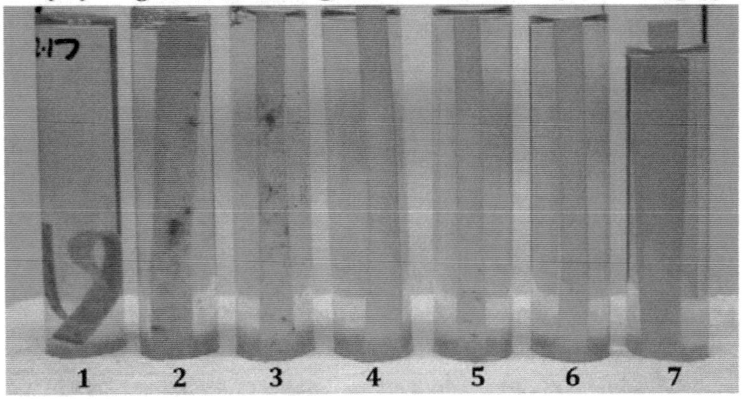

The addition of fermentable sugars delays hydrolysis of filter paper by *Piromyces sp finn.*

19

Identified Cellulolytic Enzymes are Regulated

13% of all transcripts are glucose repressed

- 30% of dockerin-fused transcripts are downregulated in response to glucose
- Transcripts from 29/54 of complete ORFs examined are glucose repressed

Regulation is likely antisense mediated

The hallmarks of transcriptional regulation are observed – how does this correspond with regulation at the protein level?

Transcripts with Dockerin Domains

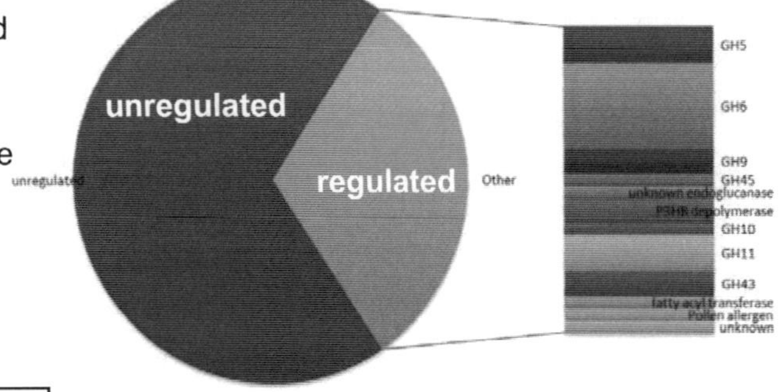

20

Kevin Solomon

Fungal Cellulosome Composition Depends on the Substrate

UCSB

- Fungal cellulosomes from *Piromyces sp finn* are readily precipitated from anaerobic cultures
- Cellulosome complexes are large, and have 10+ putative subunits
- Subunits of the complex evolve based on extracellular substrate

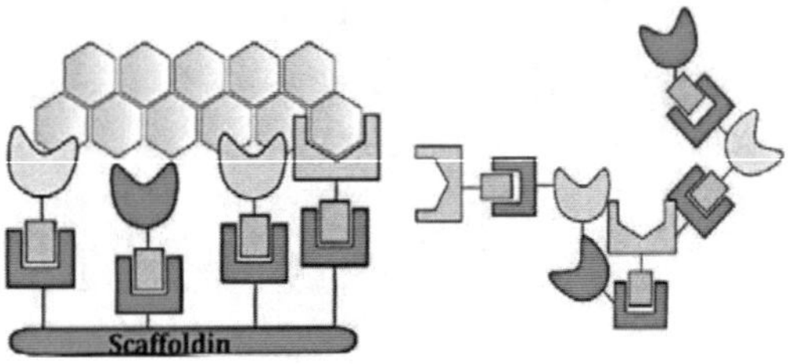

Fungal Cellulosome Architecture

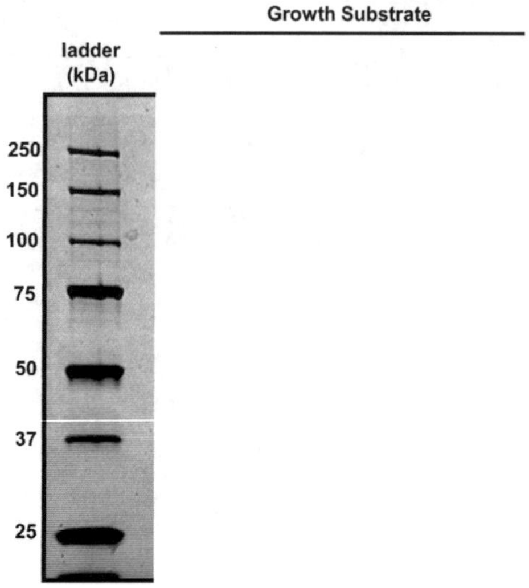

Fungal Cellulosome Components

21

Charles Haitjema

Summary and Future Outlook

UCSB

✓ We have isolated a unique, cellulosome-producing anaerobic fungus from the *Piromyces* genus

✓ Complete coding sequences for thousands of novel genes were identified from *de novo* transcriptome assembly

✓ We have observed transcriptional regulation of cellulolytic genes, including dockerin-tagged lignocellulolytic enzymes in gut fungi

Upcoming…

- Sequencing and comparison of *de novo* hits with genes within the *P. finn* genome (Broad Institute, DOE-JGI)
- Metabolic modeling and prediction of dynamic gut fungal enzyme regulation for tunable cellulosomes
- Stable genetic transformation strategies for gut fungi
- Biochemical and structural engineering of fungal cellulosomes

22

Acknowledgements

UCSB

Broad Institute:
Aviv Regev
Dawn Thompson
Diego Borges-Rivera

Durham/Aberystwyth (UK):
Mike Theodorou
Paul Wyn Jones

USDA:
Paul Weimer
Chris Odt

Verrill Farm (Concord, MA)

O'Malley Lab:
Kevin Solomon
Charles Haitjema
Nikki Schonenbach
Jen Guerrero
John Henske

UCSB (BMSE Program):
Nick Beck
Brett Cook

Kevin Solomon

Charles Haitjema

BIO-RAD

USDA

ICB

BROAD INSTITUTE

Funding and Support:

- **Company of Biologists Travel Fellowship**
- **Bio-rad**
- **U.S. Department of Agriculture (ARS-NIFA)**
- **U.S. Army - Institute for Collaborative Biotechnologies (ICB)**

"Trapping" Cellulosome Components from Gut Fungi

UCSB

Recombinant dockerins

Yeast

→

- By producing recombinant dockerins in yeast, we can develop methods to biochemically purify fungal cellulosomes
- This will help illuminate the scaffolding mechanism used by the complex

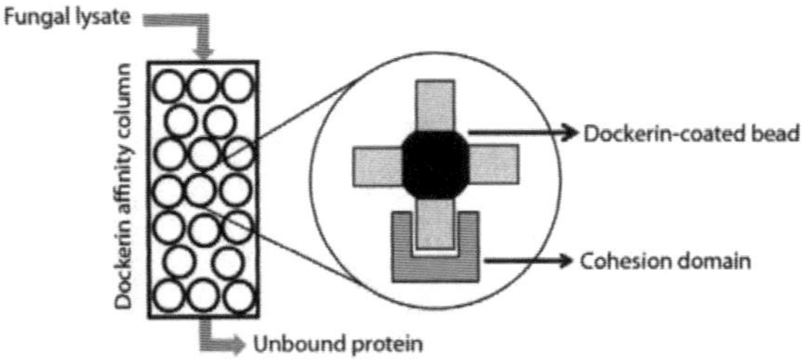

Affinity chromatography identifies interacting proteins by exploiting the cohesin-dockerin interaction

24

Intracellular Protein Delivery Using Degradable Polymeric Nanocapsules

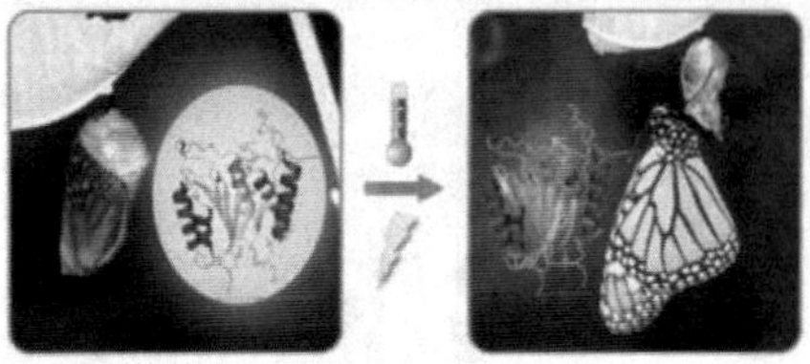

Yi Tang

Department of Chemical and Biomolecular Engineering
Department of Chemistry and Biochemistry
University of California, Los Angeles

UCLA

Research Interests in the Tang Lab

- My lab primarily tries to understand biochemistry of bioactive natural product biosynthetic pathways.

- We also engineer novel enzymes into useful biocatalysts towards the synthesis of complex pharmaceutical molecules.

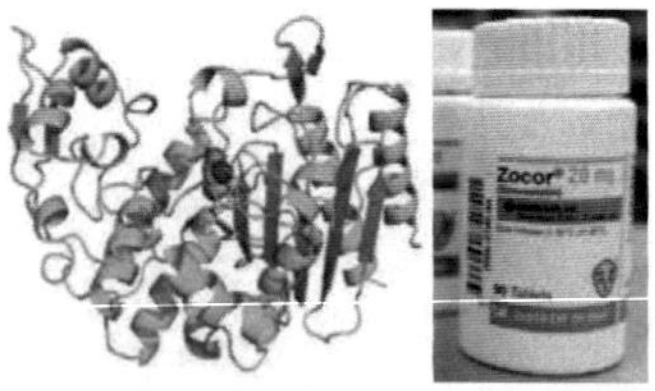

- **We have recently begun the engineering of new polymeric vehicles for intracellular delivery of protein therapeutics.**

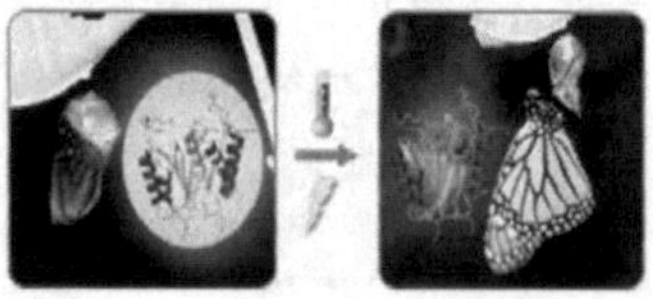

Protein-based Therapeutics

- **Advantages of Protein therapeutics**
  - **Specificity**
  - **Biocompatibility**
  - **High potency**
  - **Unique in form and function**

- **Approved Protein Therapeutics**
  - **Diabetes** (Insulin: Humulin, Novolin, Symlin, NPH…)
  - **Cancer** (Herceptin, ELSPAR, Avastin, Vectibix…)
  - **Cardiovascular** (Natrecor, Angiomax, Retavase…)
  - **Immunoregulation** (Adagen, Infergen, Intron A…)
  - **Growth regulation** (Sandostatin, Kepivance, Infuse…)

  …(~$77 billion in 2010)

However, nearly all current protein therapeutics act on extracellular targets. What about intracellular targets?

B. Leader, Q. J. Baca, D. E. Golan, ***Nat Rev Drug Discov***, **7**, 21, 2008.

Intracellular Protein Delivery
Opportunities

- **Potential Applications**
  - Catalyze intracellular reactions
  - Restore loss-of-function genetic conditions
  - Maintain normal cellular life cycles
  - Artificial control of gene expression levels
  - Imaging
  - Vaccination

Protein delivery adds functions to cells without modifications to the host genome

Intracellular Protein Delivery
Challenges

Extracellular Challenges

Intrinsically unstable (aggregation, denaturation)
Prone to proteolysis
Rapid clearance of small proteins (< 30kDa)
Elimination by immune system
Targeting specific cells

Entry Challenges

Surface charge (negative for most recombinant proteins)

Intracellular Challenges

Escape from delivery vehicles
Release into cytosol
Maintain structure and function

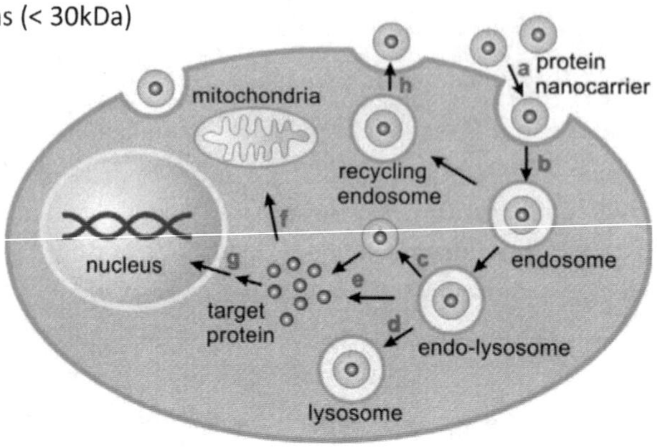

Many of the delivery challenges can be addressed using nanocarriers

Intracellular Protein Delivery –
Using Nanocarriers

Native proteins

Extracellular Challenges

Intrinsically unstable (aggregation, denaturation)
Prone to proteolysis
Rapid clearance of small proteins (< 30kDa)
Elimination by immune system
Targeting specific cells

Entry Challenges

Surface charge (negative for most recombinant proteins)

Intracellular Challenges

Escape from delivery vehicles
Release into cytosol
Maintain structure and function

Using Nanocarriers

Advantages

Increase protein stability
Shield protein from protease
Increase circulation time
Concealing immunogenic epitopes
Allow surface modification

Tailor carrier surface charge

Choice and design of vehicle is crucial

Nanocarriers for Intracellular Protein Delivery

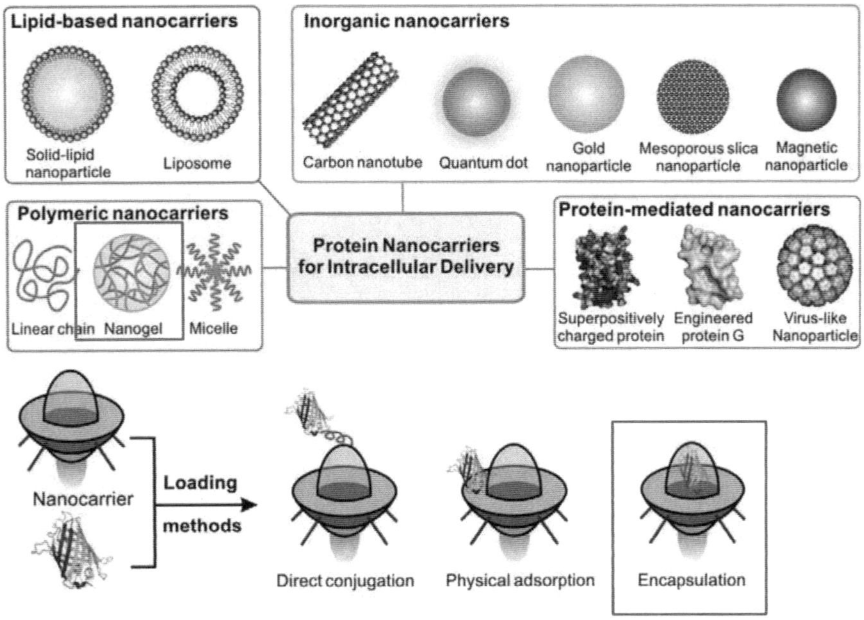

D. Peer, J. M. Karp, S. Hong, O. C. Farokhzad, R. Margalit, R. Langer, ***Nat. Nanotechnol.***, *2*, 751, **2007**.

Z. Gu, A. Biswas, M. Zhao, Y. Tang, *Chem. Soc. Rev.*, **2012**

Protein Delivery using Polymeric Nanocapsule

Outline of the Different Generations of Nanocapsules

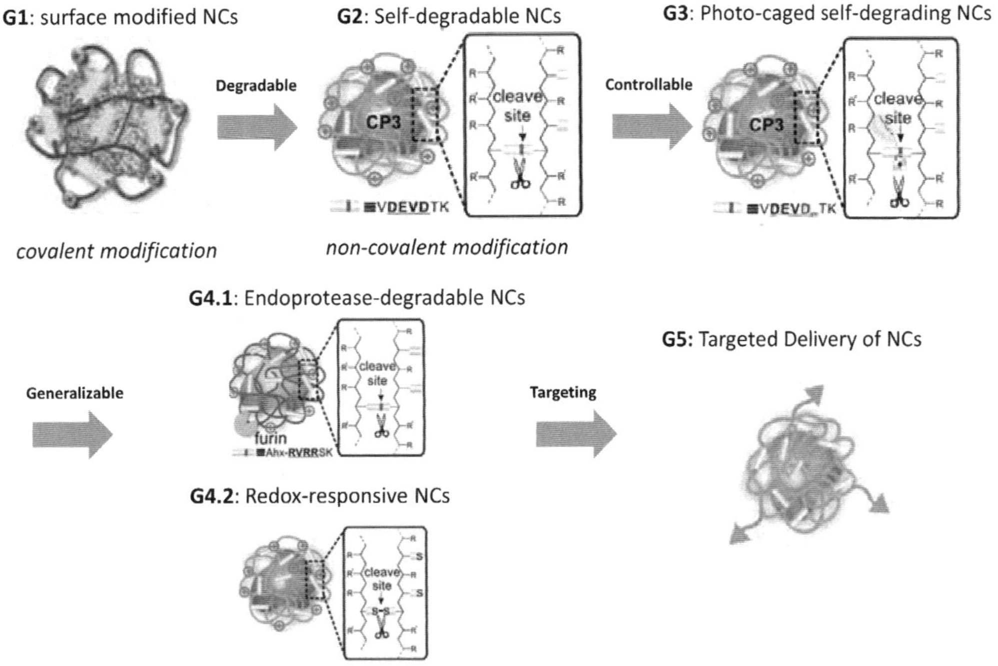

Single-Protein Based G1 Nanocapsules

G1

M. Yan, et al *Nat. Nanotechnol.*, 5 (48), **2010.**
with Prof. Y. F. Lu, Prof. T. Segura

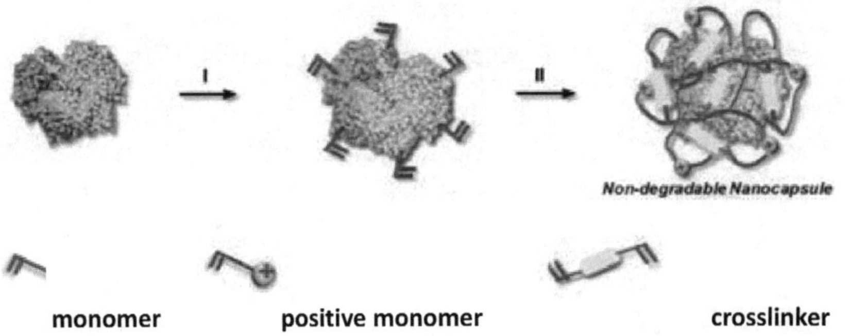

Uniform size | Increased stability | Efficient cell entry

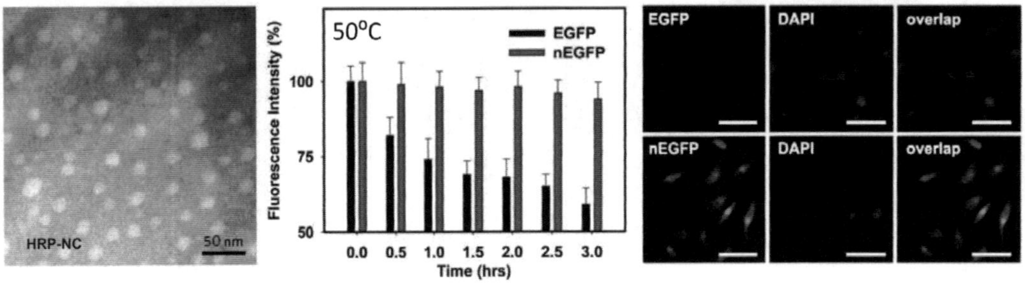

Degradable Nanocapsules

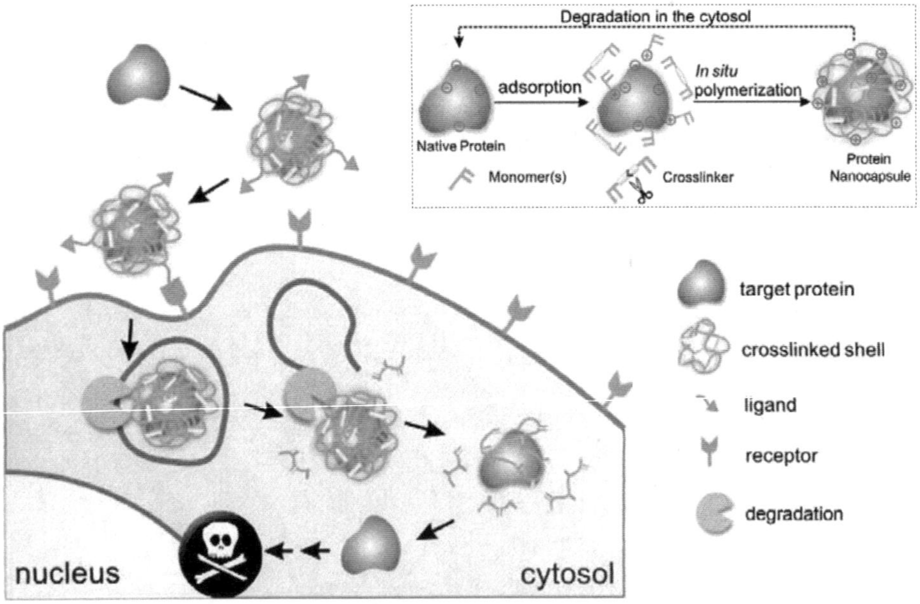

Shown here is the delivery of a protein that can cause apoptosis in cells.

Delivery of Caspase 3 (CP-3)

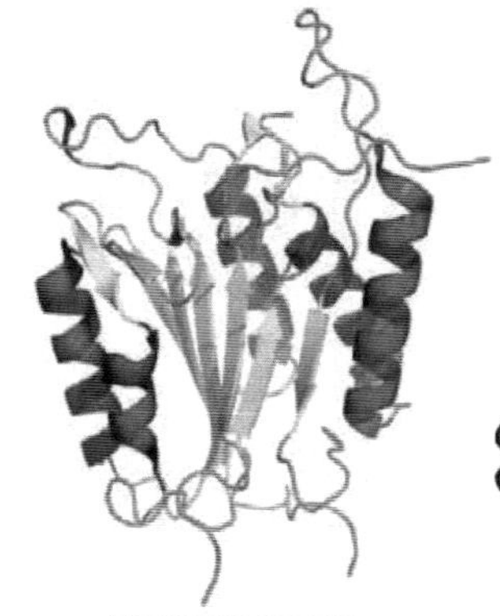

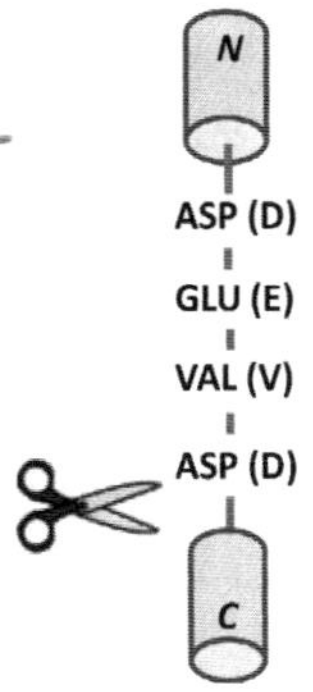

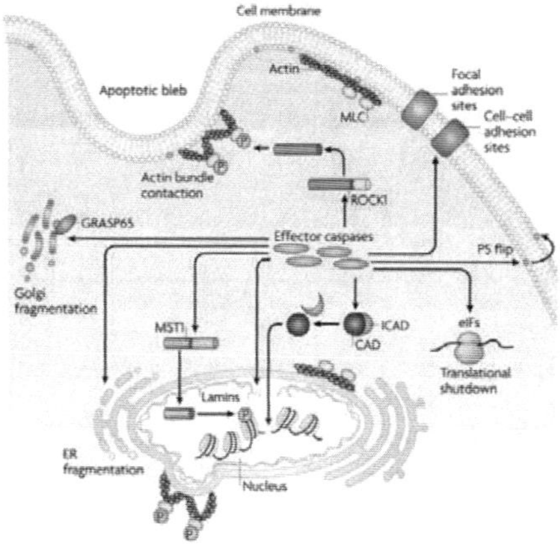

R. C. Taylor et al
Nature Reviews Molecular Cell Biology (2008)

Cytosolic delivery of CP-3 to cancer cells can rapidly induce apoptosis. To do so, CP-3 must be delivered in functional form and interact with the various macromolecular targets in the cytosol. Therefore, CP-3 must be reversibly encapsulated in a nanocapsule that can degrade.

G2

Self-Degradable G2 Nanocapsules

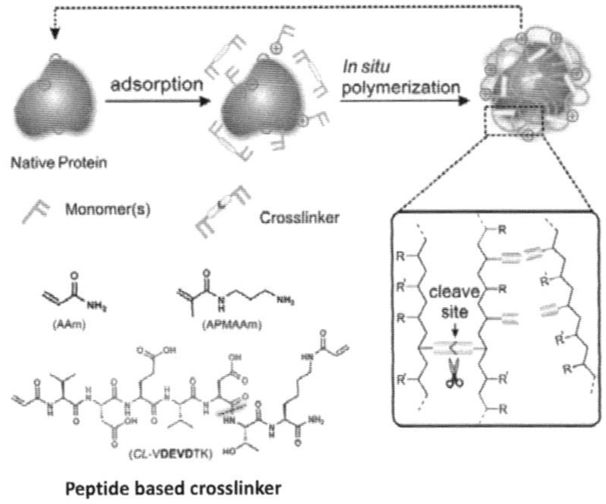

Peptide based crosslinker

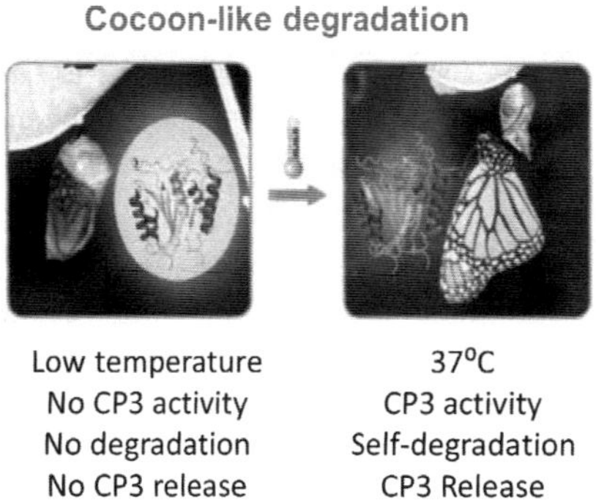

Z. Gu et al *Nano Lett.*, 12 (4533), **2009.**

Dr. Zhen Gu

G2

Characterization of NC and Self-Degradation

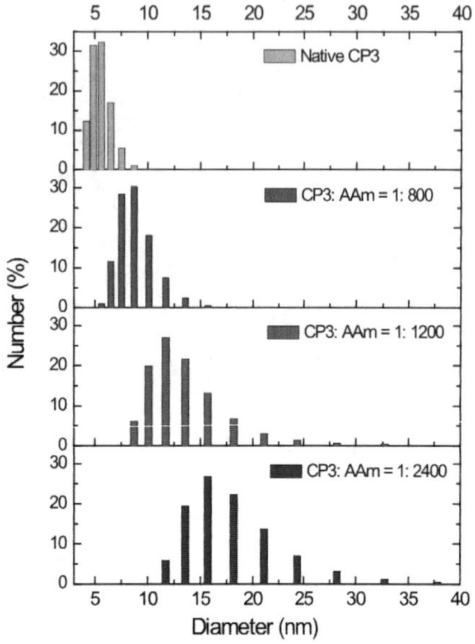

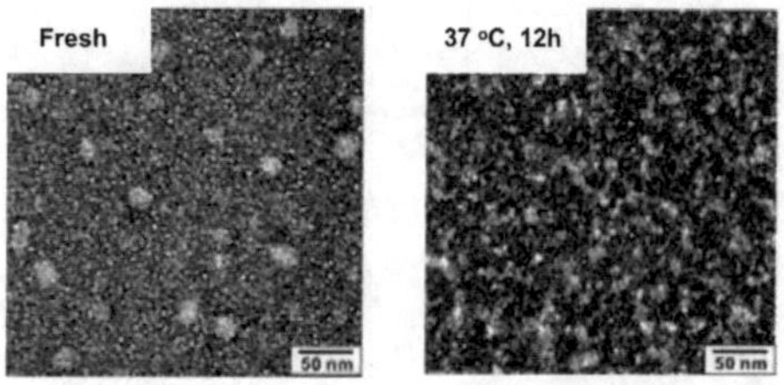

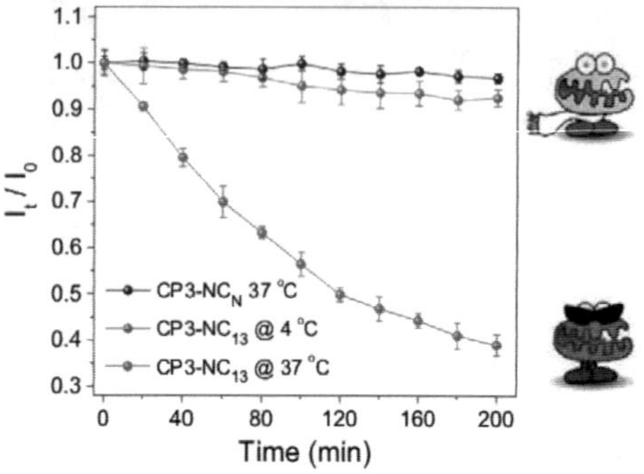

ζ-potential between 6 and 9 mV

Z. Gu et al *Nano Lett.*, 12 (4533), **2009.**

G2

Release kinetics of CP-3 from NC

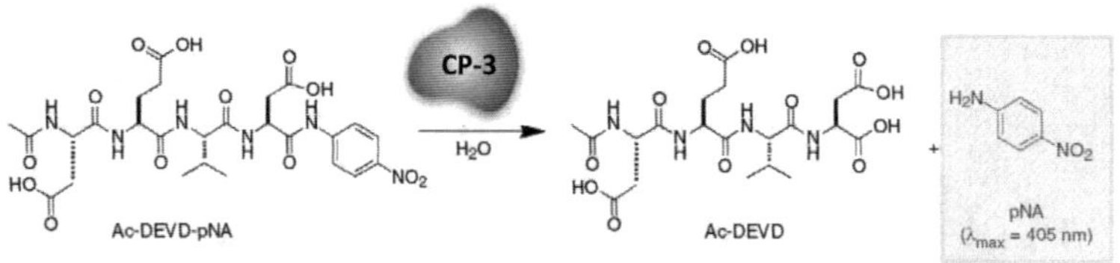

Large peptidyl substrate
Inaccessible by encapsulated CP3

Nature Protocols, **2010**, p295

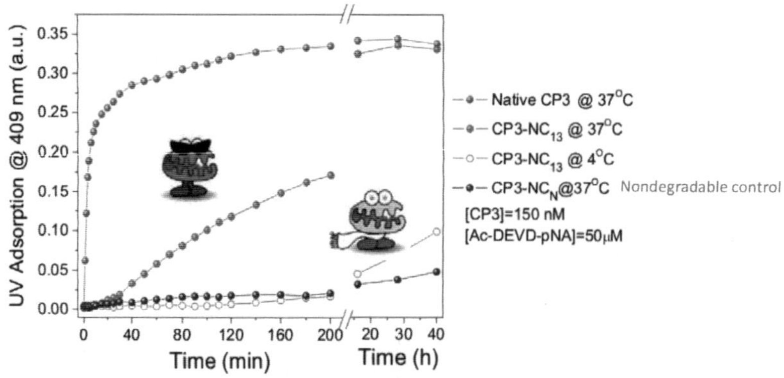

Z. Gu et al *Nano Lett.*, 12 (4533), **2009.**

Delivery of CP-3 to Cancer Cell Lines

G2

Visualization of Cargo Delivery with eGFP

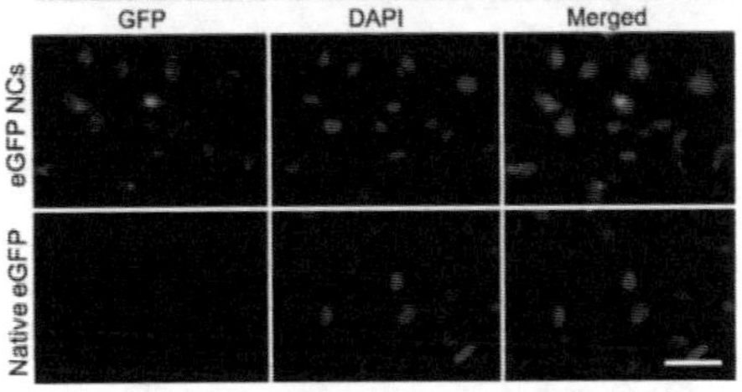

Cytotoxicity of Self-Degradable CP-3 NCs

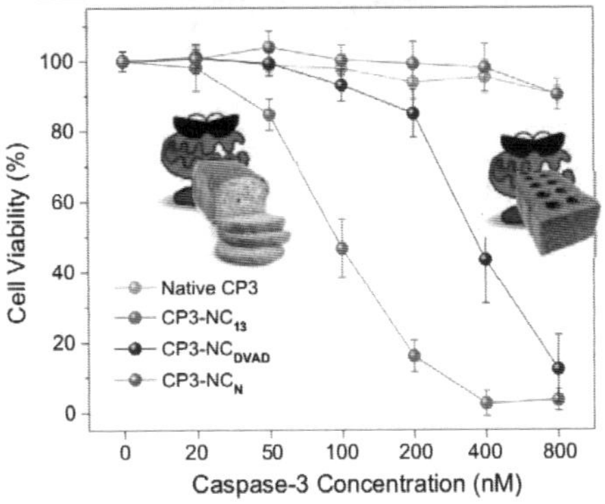

Use TUNEL assay to verify apoptosis

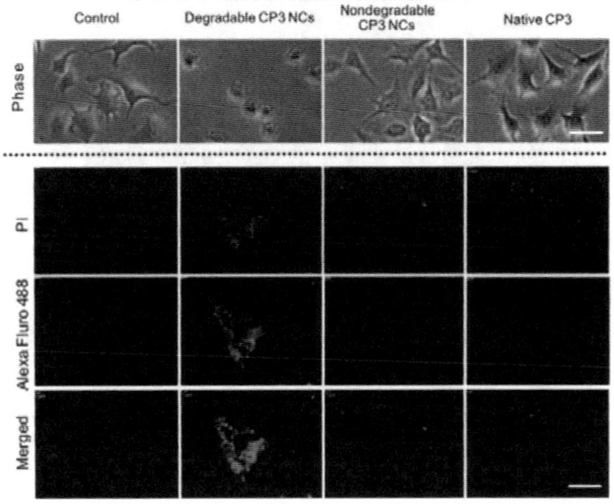

Self-degradable CP-3 NC can induce apoptosis. However, more control is needed over the timing of the degradation process.

Z. Gu et al *Nano Lett.*, 12 (4533), **2009.**

Photo-caged Self-Degrading G3 NCs

G3

Goal: to achieve spatial and temporal control of the self-degradation process

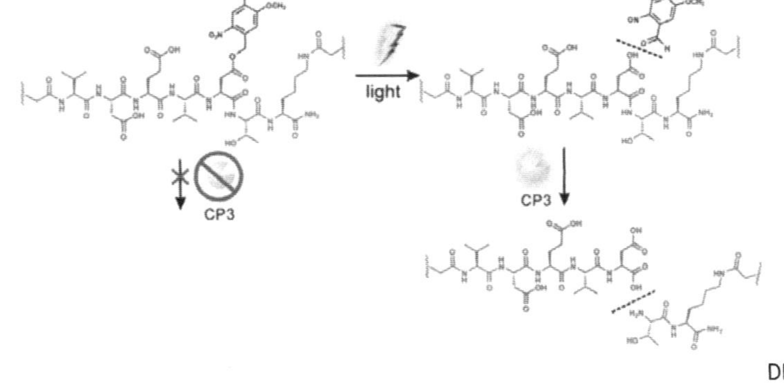

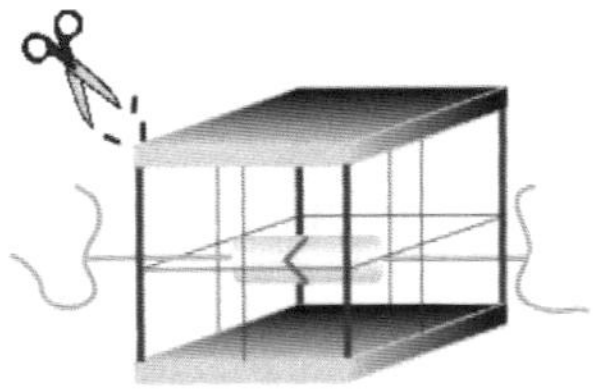

DEVD-*p*NA assay

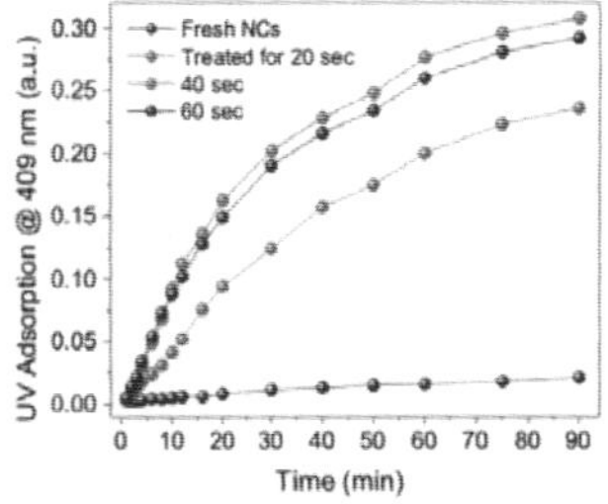

Z. Gu et al *Nano Lett.*, 12 (4533), **2009.**

Photo-caged Self-Degrading G3 NCs

G3

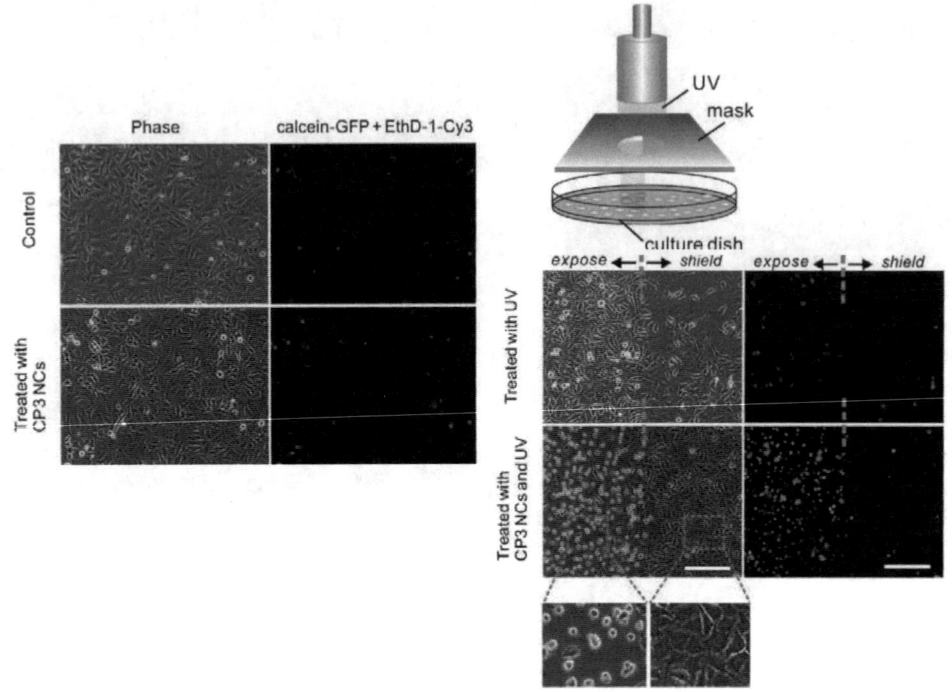

Summary of G1 – G3 Nanocapsules

- Single-protein nanocapsules synthesized from surface modified protein can penetrate the cell membrane via endocytosis.
- Encapsulated protein exhibits enhanced stability.
- Second generation nanocapsules synthesized from interfacial polymerization contains no covalent bonds between the polymeric shell and the target protein.
- Caspase 3 enclosed in the DEVD-crosslinked nanocapsule can be released via degradation of the capsule from within.
- Spatiotemporal control of the self-degradation process can be achieved via photo-caging of the peptidyl crosslinkers.

However, a more general degradation mechanism in needed to deliver a variety of protein cargo.

Strategies for Enzymatic Degradation G4.1

- Extracellular protease (Matrix Metalloproteinases)-mediated degradation of polymeric matrices have been widely used in gene therapy and drug delivery.

Cell mediated proteolytic invasion of synthetic matrix.

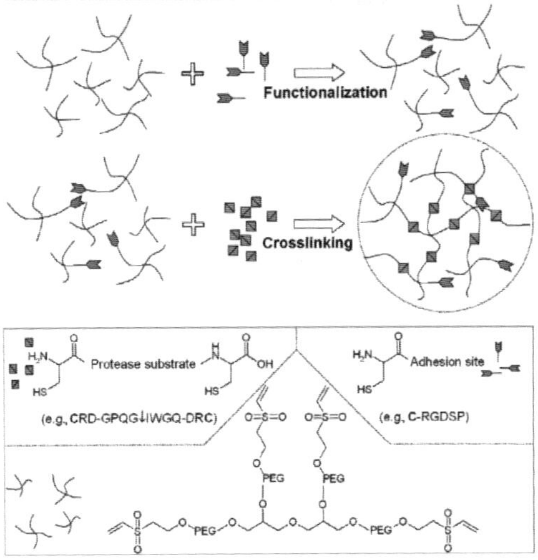

Hubbell and coworkers, *Nat. Biotechnol.* **2003**

DNA delivery from MMP degradable PEG hydrogels

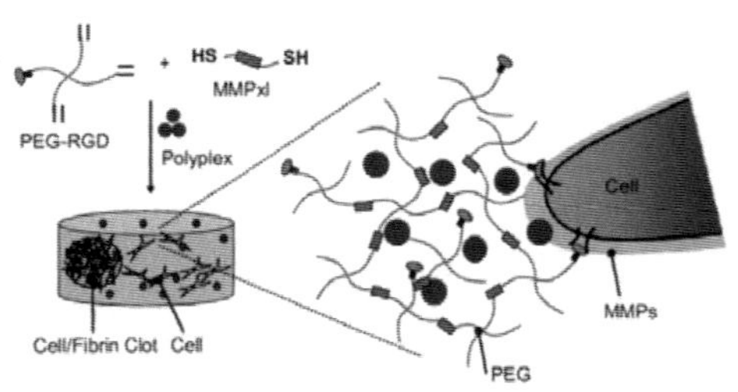

Segura and coworkers, *Biomaterials.* **2009**

To degrade the protein nanocapsules for cytosolic delivery, the actions of ***intracellular*** proteases can be explored.

Endoprotease-degradable G4.1 Nanocapsules G4.1

Furin – 53 kDa Proprotein convertase

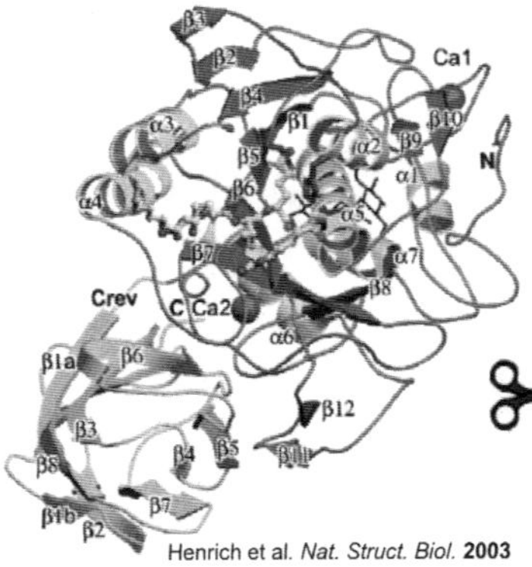

Henrich et al. *Nat. Struct. Biol.* **2003**

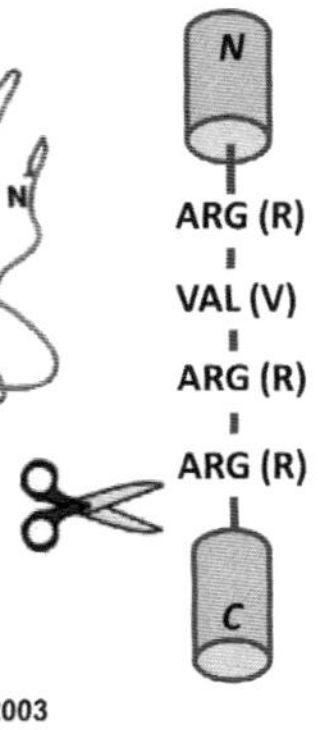

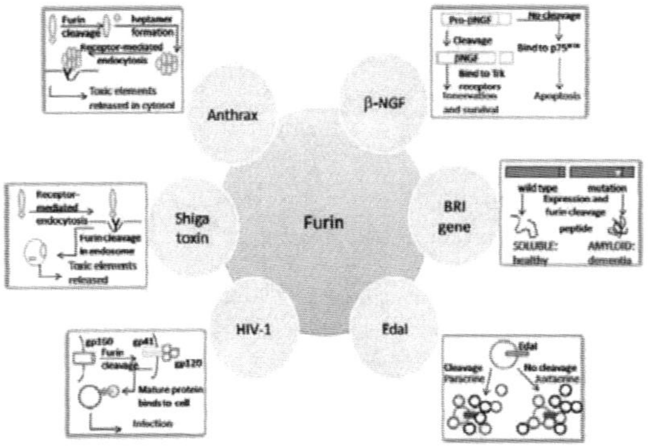

Thomas et al. *Nat. Rev. Mol. Cell. Biol.* **2002**

- Required for the maturation of various cellular proteins.
- Found in various intracellular locations, including cell membrane, endosomal compartments, and trans-Golgi network.
- Increased furin expression in breast, ovarian, brain, lung and neck cancers.

Anuradha Biswas

Biomimetic Inspiration

Furin processes both cell-encoded substrates and foreign material.

PNAS

Cleavage of the papillomavirus minor capsid protein, L2, at a furin consensus site is necessary for infection

Rebecca M. Richards, Douglas R. Lowy, John T. Schiller, and Patricia M. Day* *PNAS.* **2006**

Laboratory of Cellular Oncology, Center for Cancer Research, National Cancer Institute, National Institutes of Health, Bethesda, MD 20892

Edited by Peter M. Howley, Harvard Medical School, Boston, MA, and approved December 15, 2005 (received for review October 7, 2005)

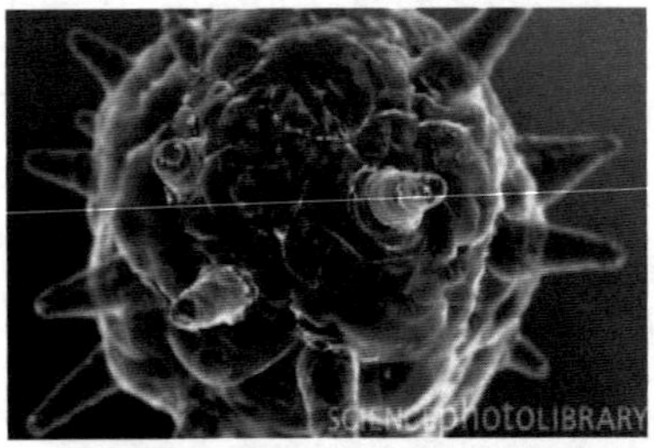

Type	Sequence
BPV1	MSARK**RVKR**ASA...
HPV16	MRHKRSAK**RTKR**ASA...
CRPV	MVARS**RKRR**AAP...
HPV4	MQSLS**RRKR**DSV...
HPV5	MARAK**RVKR**DSV...
HPV6	MAHSRAR**RRKR**ASA...
HPV18	MVSHRAA**RRKR**ASV...

Furin-mediated cleavage of papillomaviruses is necessary for dissociation of capsid, release of viral DNA and infection. Natural role of furin facilitated release of foreign cargo.

Furin-Mediated Release of Protein Cargo

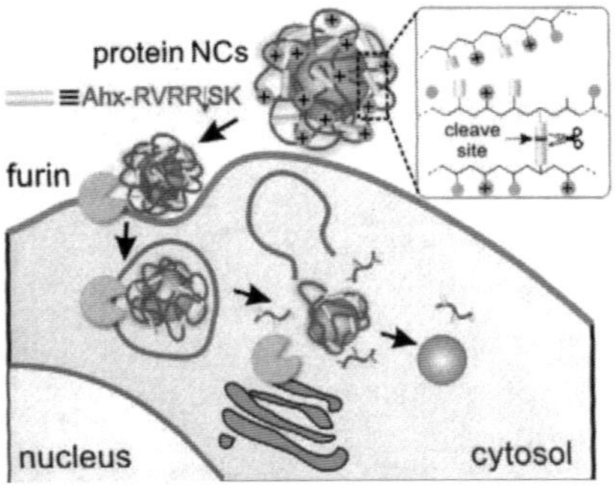

Monomers

Crosslinker

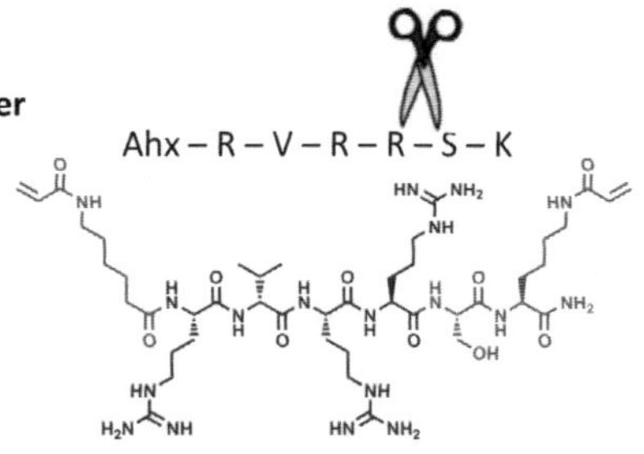

Size of eGFP nanocapsule crosslinked with RVRR linker
ζ-potential ~ 7 mV

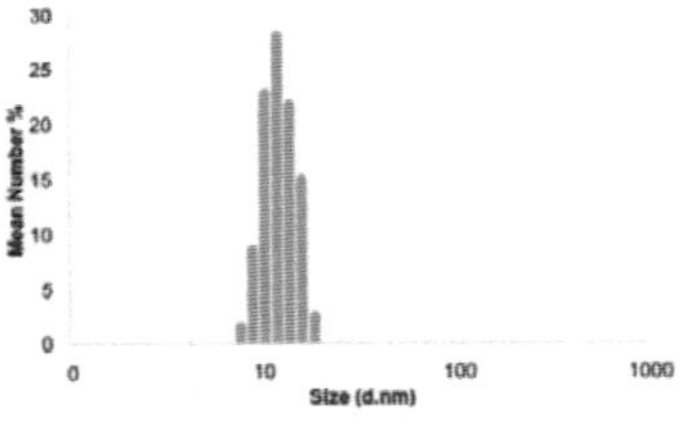

Before adding furin | 10 hr after adding furin

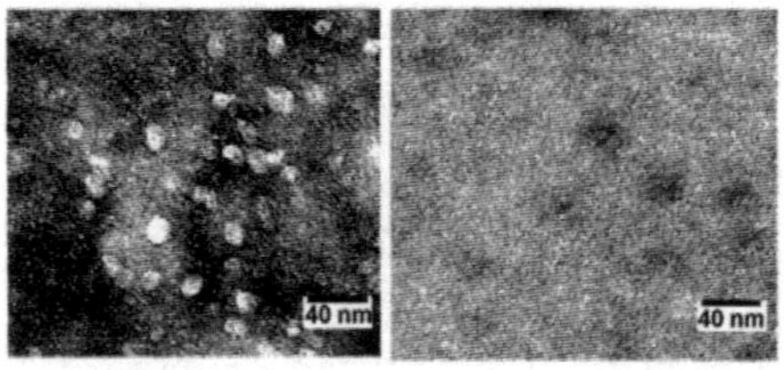

Biswas, A. et al, *ACS Nano,* **2011**.

G4.1

Furin-Mediated Release of Protein Cargo

Quantification of 2.5 nmol eGFP release from NCs using ELISA

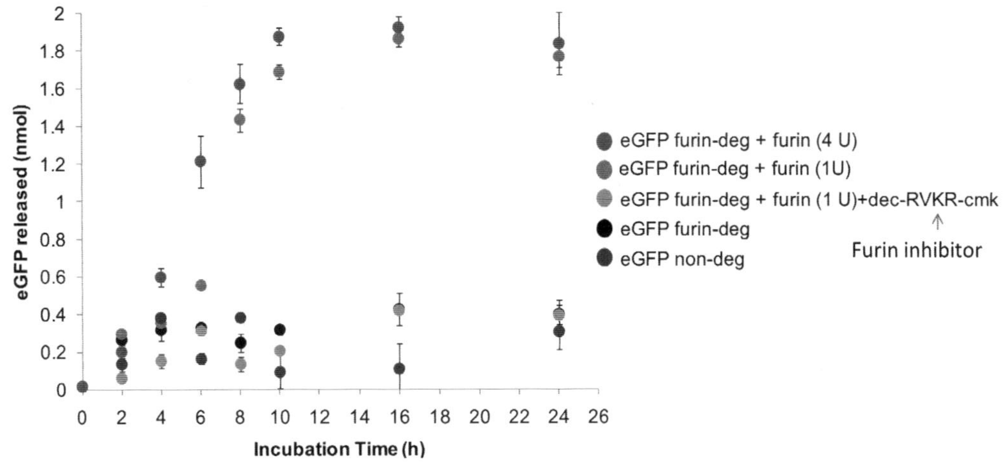

Release of native eGFP from NCs is specifically dependent on enzymatic activities of furin.

Biswas, A. et al, *ACS Nano,* **2011**.

Furin-Mediated Intracellular Protein Delivery

G4.1

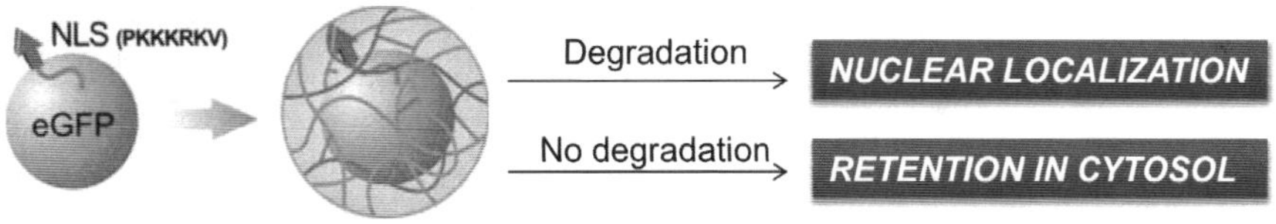

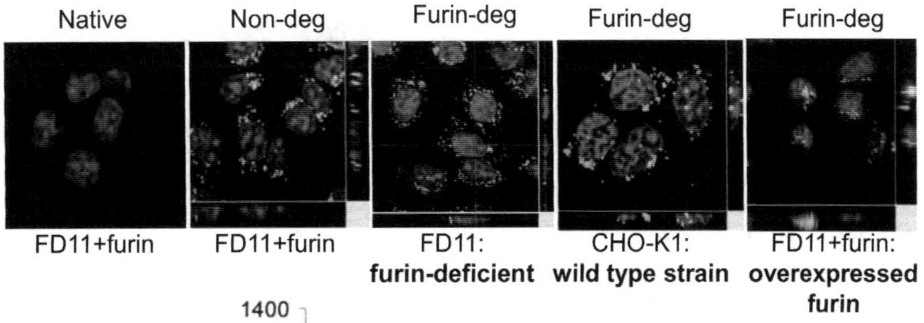

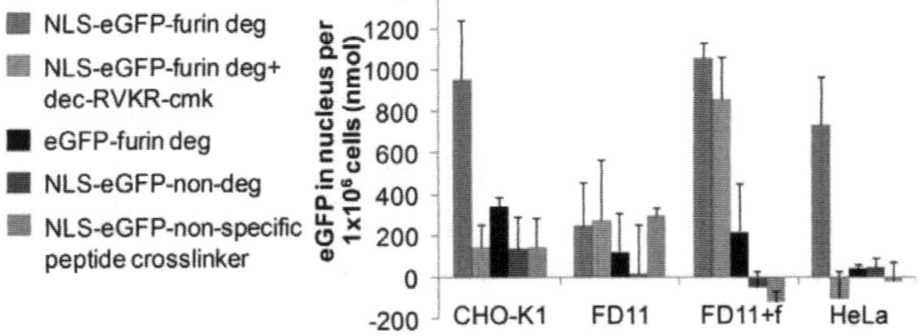

Both intracellular furin and degradable crosslinker are necessary for nuclear delivery

G4.1

Furin-mediated delivery to a variety of cell lines

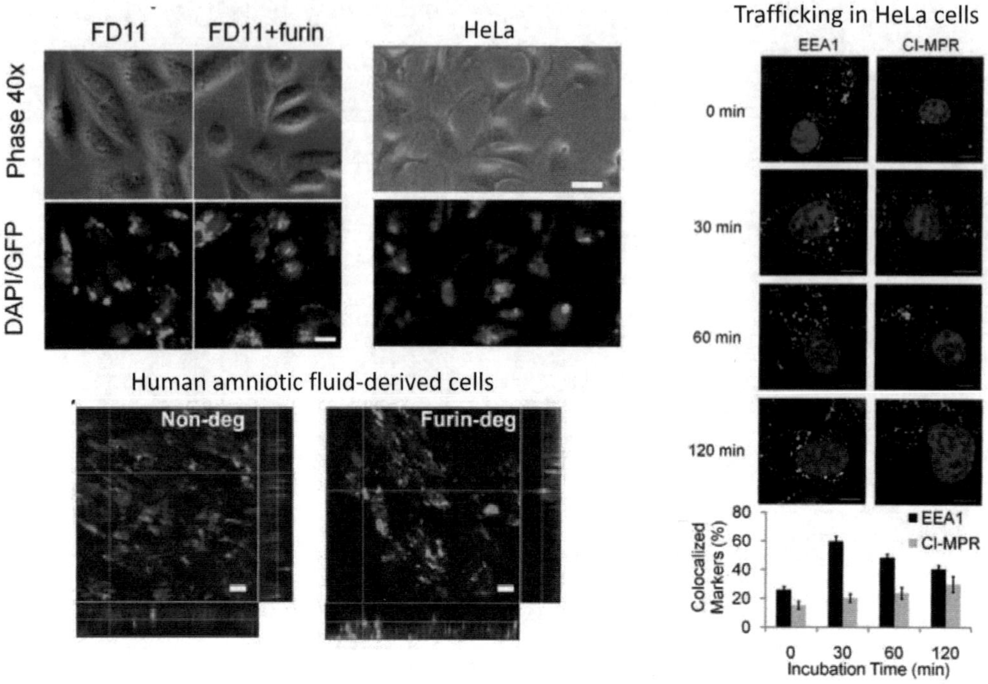

Strategy for Chemo-Degradation

G4.2

- The difference in the redox environment between extracellular (oxidizing) and intracellular (reducing) can also be explored for degradation of the polymeric shell.
- The high intracellular concentrations of glutathione (GSH) can rapidly trigger reduction of disulfide (S-S) crosslinked matrices.

Gene delivery from redox degradable polyplexes

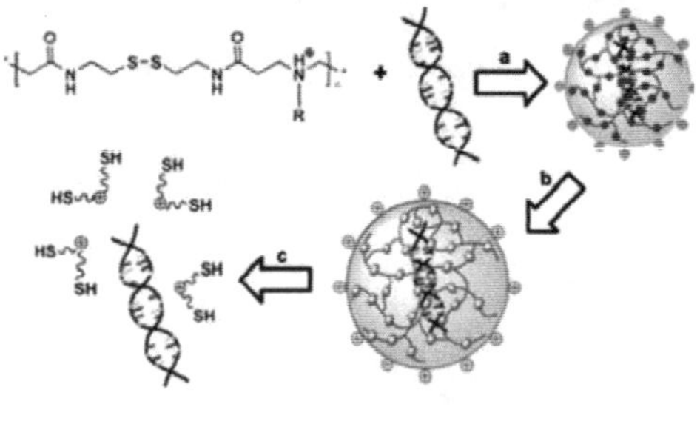

Engbersen and coworkers, *Bioconjug. Chem.* **2007**

Disulfide-linked siRNA multimers for gene silencing

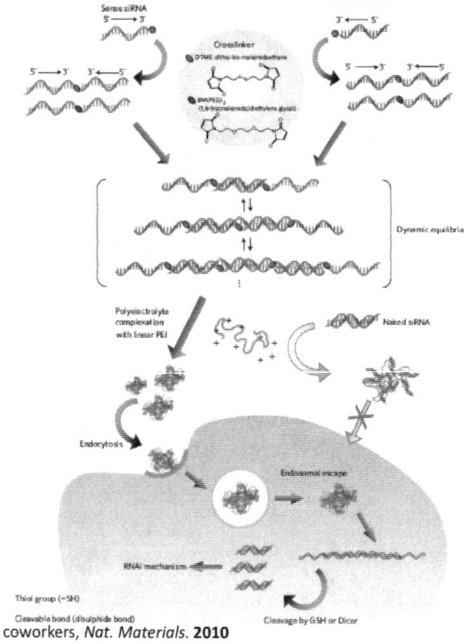

Park and coworkers, *Nat. Materials.* **2010**

Redox-Responsive G4.2 Nanocapsules

G4.2

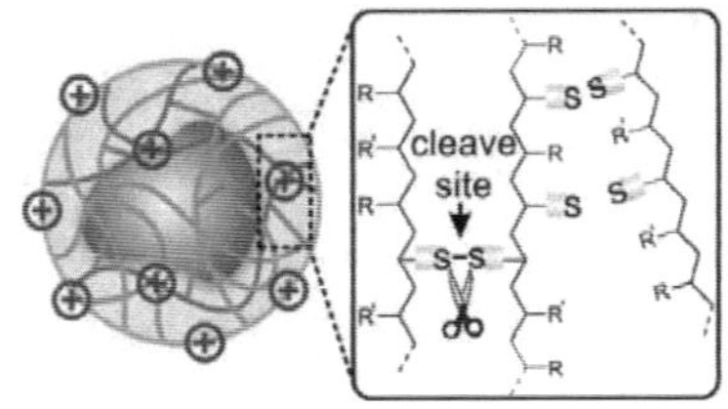

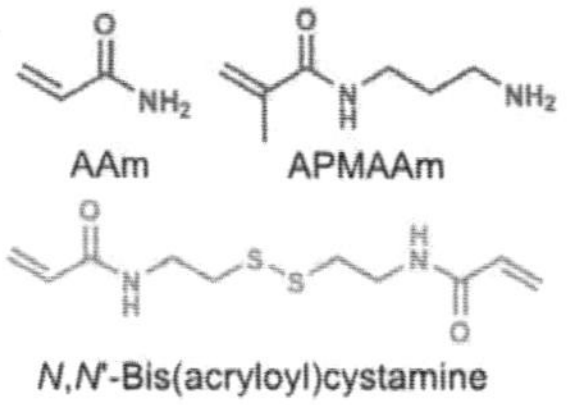

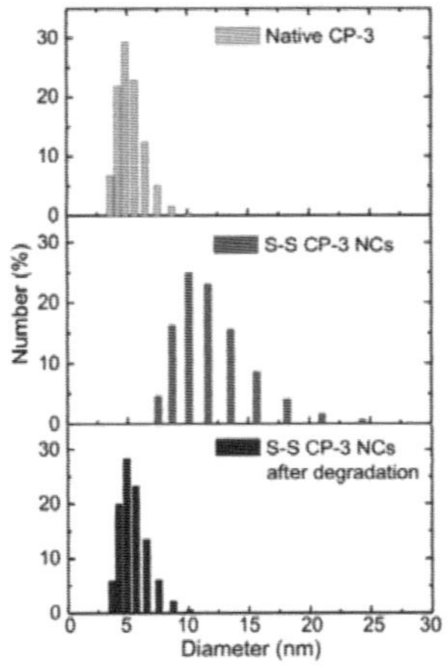

Before GSH Treatment   After GSH Treatment

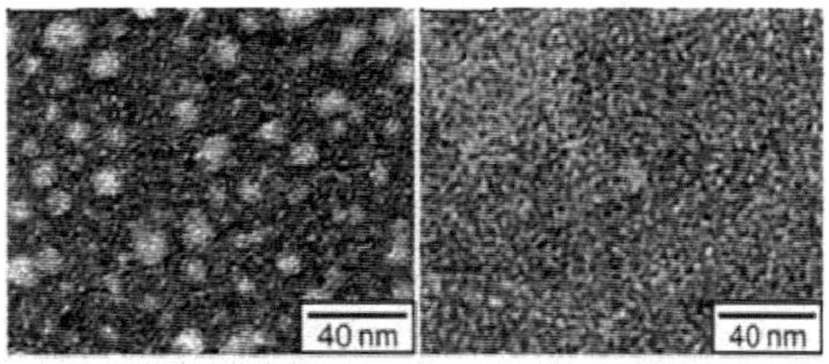

ζ-potential ~ 4 mV   2 mM GSH, 2 hr, 37°C

Zhao, et al, *Biomaterials,* **2011**

Muxun Zhao

Redox-Responsive G4.2 Nanocapsules

G4.2

Time course of GSH-mediated degradation

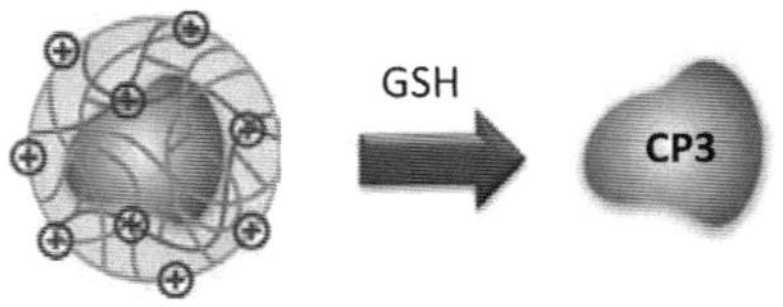

CP3 release in the presence of GSH

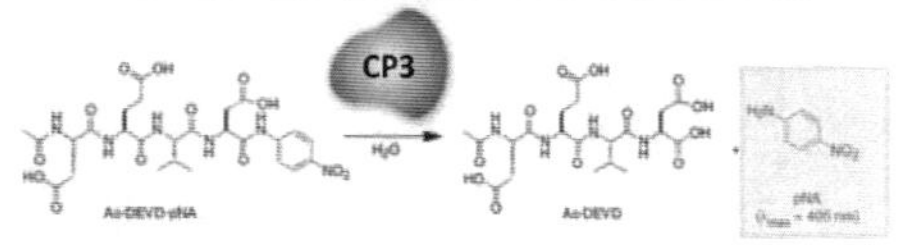

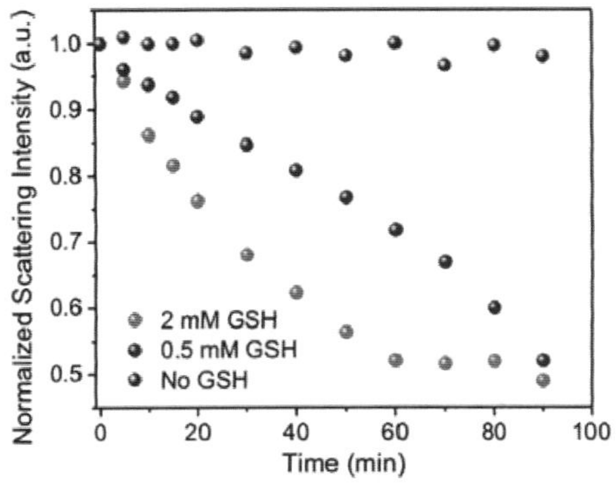

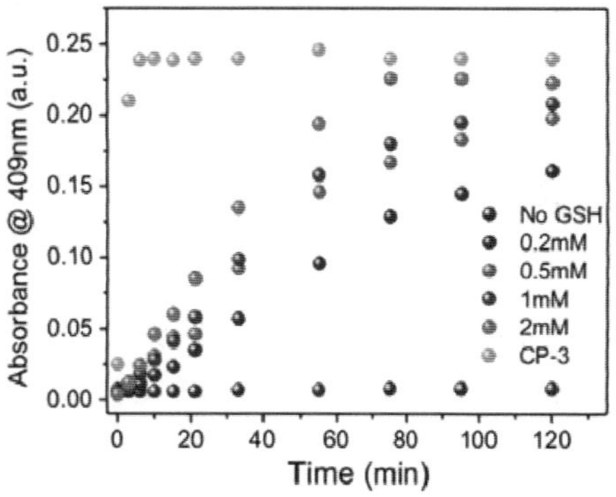

Zhao, et al, *Biomaterials,* **2011**

Endocytosis of Redox-Responsive Nanocapsules

G4.2

S-S eGFP Nanocapsule Internalization

S-S eGFP NC Trafficking in HeLa

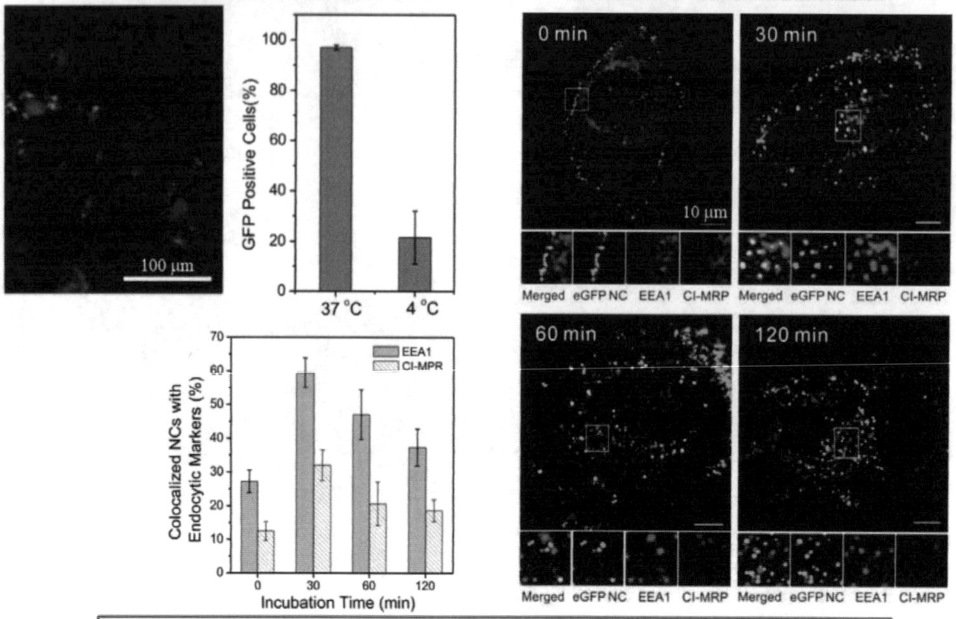

Some of the NCs have been released into cytosol after two hours of internalization

CP3 Delivery using Redox-Responsive NCs

G4.2

S-S NC with Rhodamine-CP3

TUNEL Assay to Verify Apoptosis

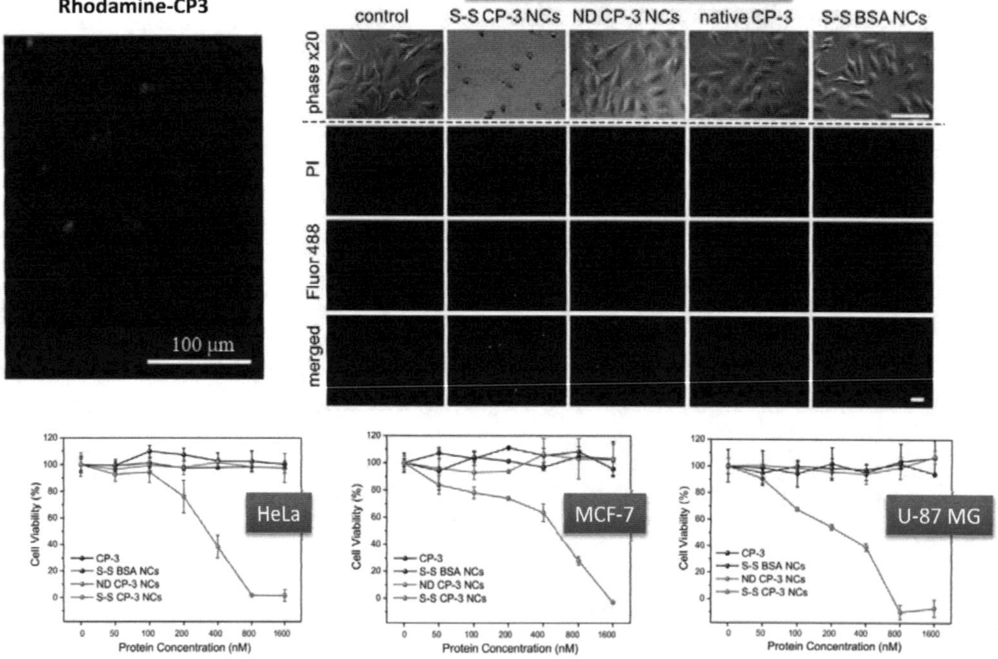

G4

Application: Apoptin

Apoptin induces apoptosis selectively in tumor cells, but not in normal cells

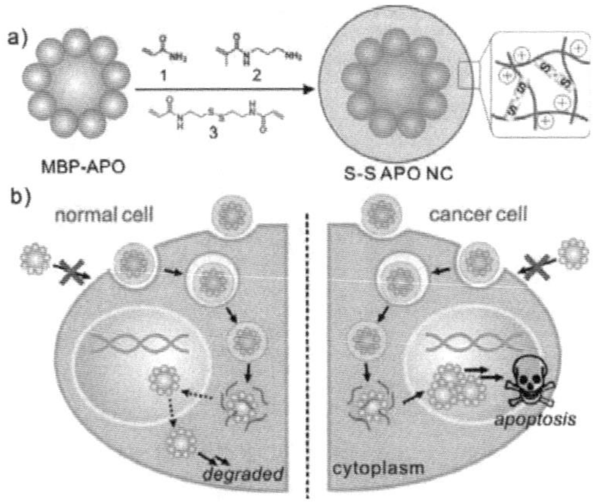

Native MBP-APO (2.4 MDa Complex)

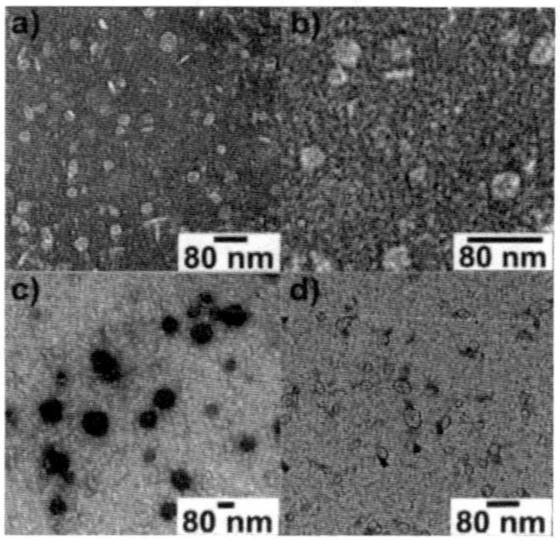

After encapsulation After degradation

Potential antitumor therapeutic if the correct delivery vehicle can be designed.

Zhao, M. et al, *Nano Today*, **2013** in press

G4

Delivery of Apoptin to Selectively Kill Tumor Cells

Nanocapsule delivered-apoptin localization

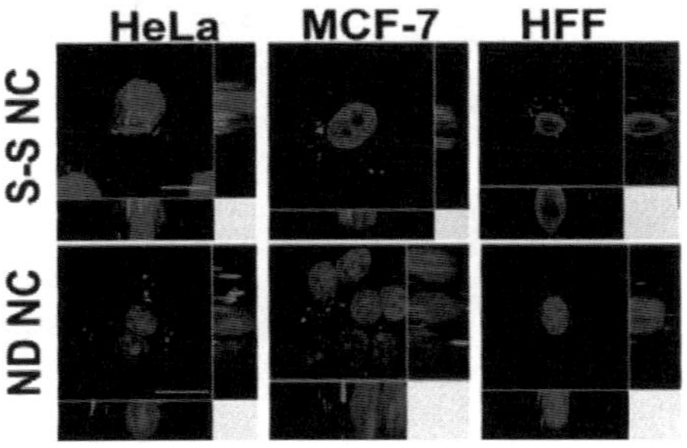

Nanocapsule can slow down tumor growth in mouse MCF-7 breast cancer xenografts.

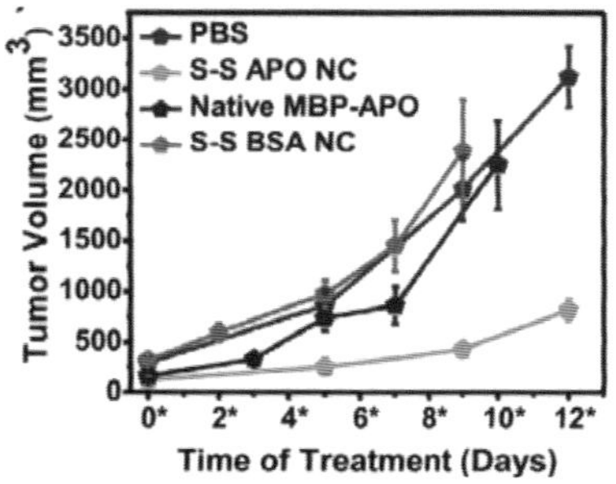

Normal Cell (HFF) Cancer Cell (MCF-7)

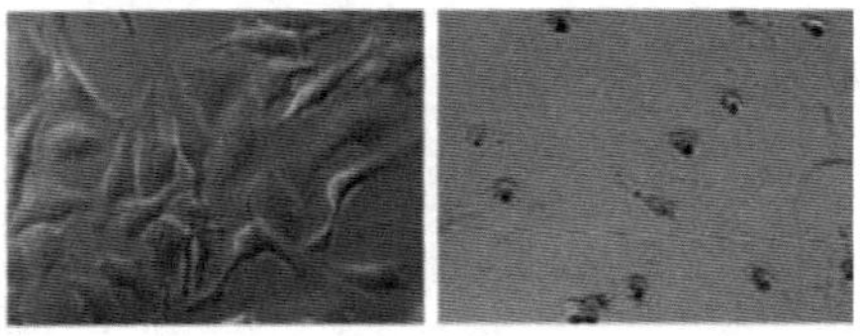

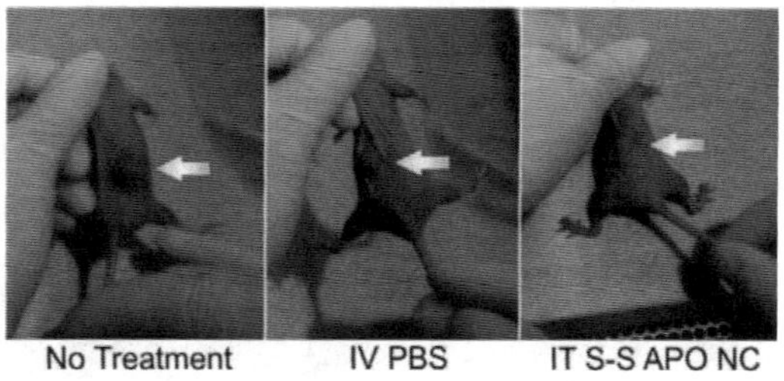

No Treatment IV PBS IT S-S APO NC

Zhao, M. et al, *Nano Today*, **2013** in press

G4

Application: Transcription Factors

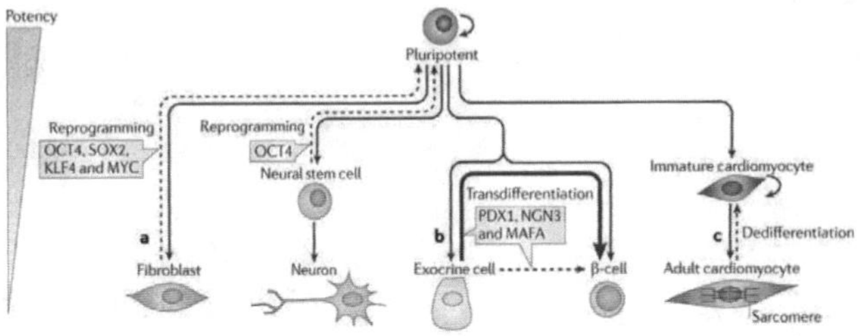

Jopling et al. *Nature Reviews Molecular Cell Biology*, **2011**

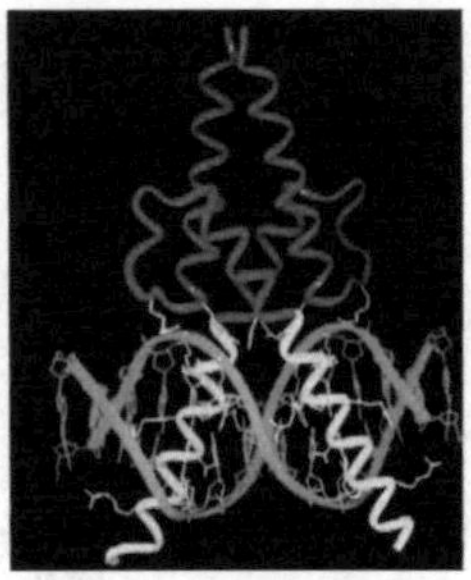

Red/yellow: TF
Blue: DNA

- Regenerative medicine aims to help repair diseased or damaged tissues by replacing affected cells with healthy functional cells
- Transcription factors are driving forces for directing cellular fate
- Effective TF delivery requires efficient nuclear delivery of proteins in active form

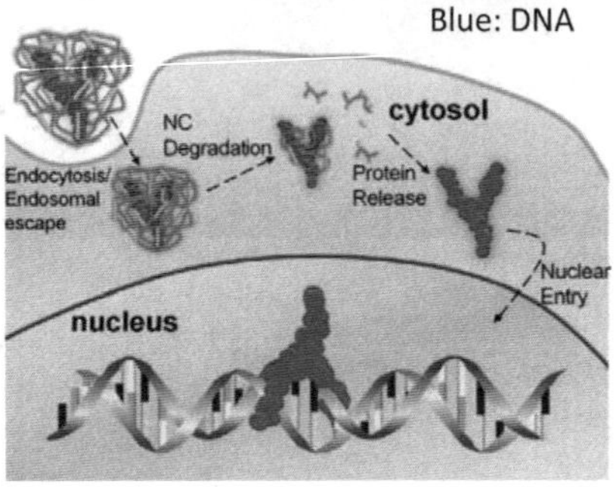

Biswas, M. et al, *Biomaterials*, **2012**

G4

Delivery of MyoD

MyoD is one of myogenic regulatory factors which act sequentially in myogenic differentiation

Key role of MyoD is to commit mesoderm cells to a skeletal lineage, and then to regulate that process

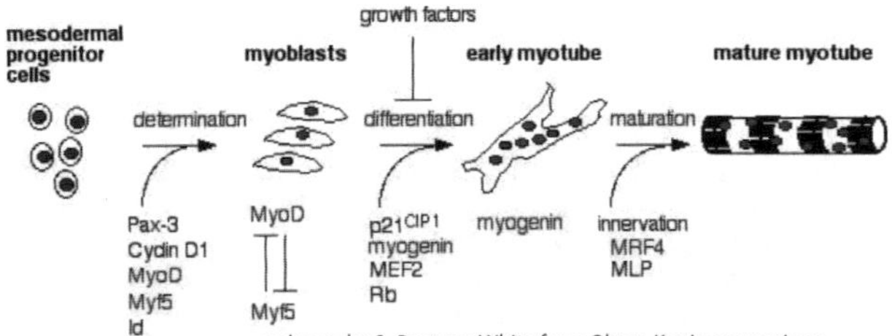

redrawn by S. Pearson-White from Olson, Konieczny reviews

MyoD nanocapsule drives myotube formation

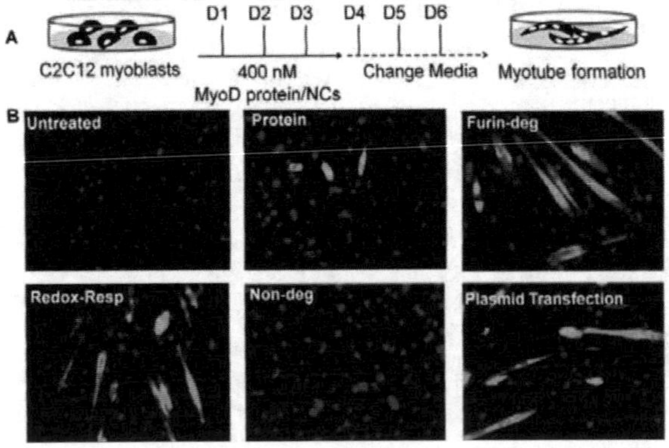

Green: MyHC Ab; Blue: DAPI-stained nuclei

MyoD nanocapsule is resistant to protease

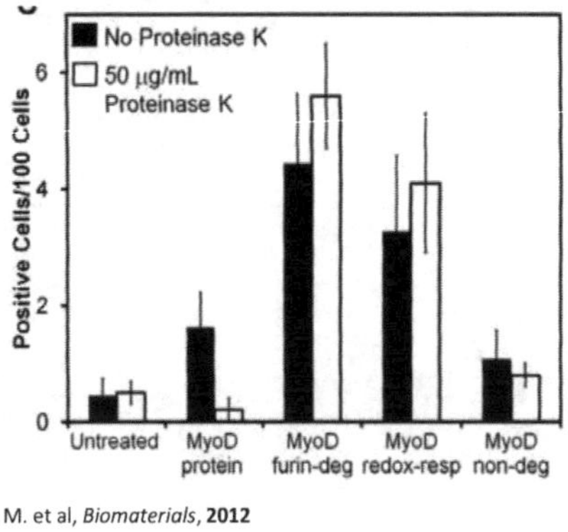

Biswas, M. et al, *Biomaterials*, **2012**

Summary of G4 Nanocapsules

- Nanocapsules have been synthesized to degrade in response to intracellular protease Furin.
- NLS-tagged protein cargo, upon release from the nanocapsule, can enter the nucleus of the cell. This can therefore be used as a method to deliver transcription factors.
- Disulfide-containing nanocapsules can be rapidly degraded in the more reducing (GSH) environment of the cytosol.
- Functional proteins can be released to interact with macromolecular targets, as demonstrated in CP3 (apoptosis).
- Both furin- and redox-responsive nanocapsules can escape from the endosomal compartments and be found in the cytosol within 2 hours of internalization.

Future Directions

STRUCTURE ⇄ FUNCTION

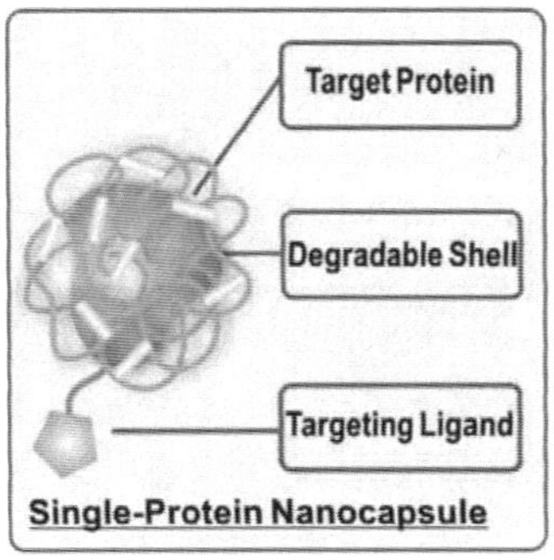

Acknowledgement

Current Members

- Anuradha Biswas
- Muxun Zhao

Former Members

- Dr. Zhen Gu

Collaborators

- Prof. Guoping Fan (UCLA)
- Prof. Pin Wang (USC)

Funding Sources

Defense Threat Reduction Agency (DTRA)

Department of Defense BCRP

UCLA Broad Stem Cell Research Center

Carved by Caspase 3
$\Phi = 600\ \mu m$

Microelectrode Array-Based Neurotoxin Biosensor with Chick Embryo CNS Neurons

Yan Kuang[1], Mauris DeSilva[2], Bin Xu[3], Tingfei Xi[4], and Bruce Z. Gao[1]

[1]Department of Bioengineering, Clemson University, Clemson, SC
[2]Naval Medical Research Unit San Antonio, Fort Sam Houston, TX
[3]National Engineering Laboratory for Regenerative Medical Implantable Devices, Guangzhou, China
[4]Center for Biomedical Materials and Tissue Engineering, Peking University, Beijing, China

The 4th International Conference on Biomolecular Engineering
January 13-16, 2013, Fort Lauderdale, FL

1. INTRODUCTION

1.1 What is urgently needed in the area of environmental neurotoxin assessment?

1.2 What is the current trend, and what is the gap?

1.3 What are we doing to fill the gap?

1.1 What is urgently needed in the area of environmental neurotoxin assessment?

- Every year, > 15,000 chemicals produced in quantities greater than 10,000 pounds per year;
- Every year, > 2,800 chemicals produced in quantities greater than one million pounds;
- Only 7% have been evaluated for their potential effects on human development.
- Every year, 2,000 to 3,000 new chemicals are reported to the U.S. Environmental Protection Agency for review to be manufactured.

⟷ correlation

- 360,000 U.S. children (1 in every 200 U.S. children) suffer from developmental or neurological deficits caused by exposure to known toxic substances.
- Autism is the fastest growing developmental disorder, increasing at a rate of 10 to 17 percent annually according to the Autism Society of America.
- Surveys conducted in California indicate an almost 210% increase in the number of cases of autism in children over the past 10 years.
- It is estimated that "3.4 of every 1,000 children ages 3-10" has an autism spectrum disorder.

⇓

Since the overwhelming majority of the 80,000 chemicals in commerce has never been tested for developmental and neurological effects, the number of children affected by all developmental and neurological toxins is probably much higher.

1.1 What is urgently needed in the area of environmental neurotoxin assessment?

Shortcomings of conventional neurotoxicity assessment:

- animal-intensive
- time-consuming
- not designed to detect the functional changes that often take place prior to the changes in structure

The development of novel neurotoxicity assay tools requires features that are

- rapid
- screening-based
- in vitro
- neurophysiological and functional
- highly sensitive
- cost-effective

1.2 What is the current trend, and what is the gap?

Trend | Biosensor: cultured neuronal network -- in vitro
Assessment: microelectrode array (MEA)-- functional & rapid

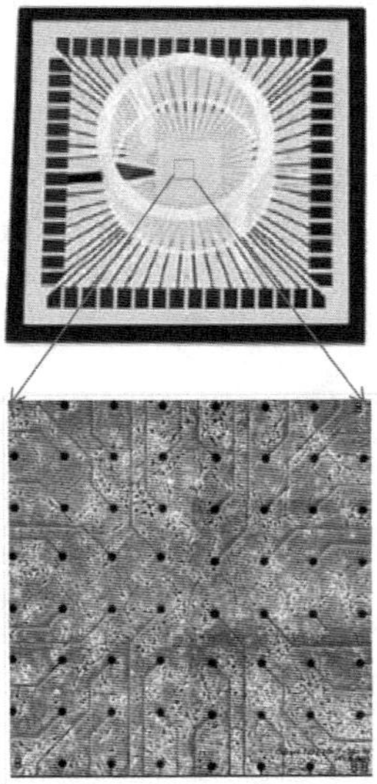

Figure 2: Rat neurons with micro-electrode array. (Photo courtesy of Steve Potter, Georgia Tech Laboratory for Neuroengineering)

1.4 x 1.4 mm^2

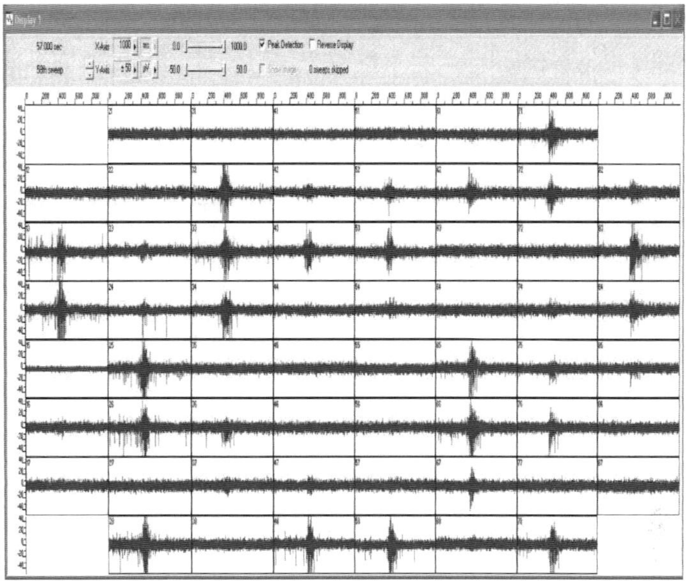

1.2 What is the current trend, and what is the gap?

The great power of MEA lies in its features:

- Sensitive measurement of network electrical activities
- Both temporal and spatial MEA traces
- Online and off-line computerized signal analysis
- Rapid high throughput of end-point data

However, MEA-based neurotoxin sensors are

- Made of rat/mouse cortical/hippocampal neurons
- Mammalian
- Still costly

Can chick neurons be utilized as neurotoxin biosensors?

- Abundant source
- Much cheaper
- Appropriate for rapid neurotoxin screening**?** **Gap**

1.3 A necessity to explore MEA-based chick biosensors

Possibility – based on shared common basic neurophysiological features between mammalian and avian

- excitability
- conductivity
- synaptic transmission
- synaptic plasticity
- certain sequential gene expression conserved in mammalian and avian

A potential concern is whether a chick forebrain neuron (FBN) network can develop to have matured functional synapses since cultured chick FBNs have a short life span in contrast to mammalian rats and mice

1.3 A necessity to explore MEA-based chick biosensors

Feasibility based on the following:

- Louis et al. (1981) paper (biochemical and TEM) demonstrated that chick FBN formed synapses in culture much faster than their speed in vivo. This is interesting.
- Preliminary MEA data from our lab on chick FBN confirmed the feasibility of using chick FBN to develop neurotoxin biosensors by characterizing the MEA physiology of chick FBN network in culture.

1.3 A necessity to explore MEA-based chick biosensors

Characterization Process

- Current characterization process:
  1) the life span of the chick FBN network in culture
  2) the time course of its spontaneous spiking activity (SSA)
  3) the spatial distribution of its SSA
  4) the patterns of SSA in comparison with its mammalian counterpart, rats or mice. (**Here we report these results.**)
- Next step: Test how a chick FBN network in culture responds to a series of classic agonists and antagonists glutamate receptors and GABA receptors (such as NMDA vs. AP5; Bicuculline vs. Muscimol) and blockers of voltage-gated channels (Verapamil and TTX, etc.) in comparison with results from rats or mice.
- Final step: Test how chick FBN network in culture responds to various known neurotoxins and potential neurotoxic chemicals.

2. Materials and Methods

- Freshly dissociatedE8 chick FBNs plated on an MEA pre-coated with PEI at a high density of 3000 cells/mm^2 without any treatment to inhibit glial growth
- Medium 199 (11043, Gibco) supplemented with 10% FBS and 2% B27, which is half changed once or twice a week
- MEA with a Teflon® lid in a dry incubator at 37 °C and 5% CO_2 (Potter, 2001)
- Ten-min measurement of the course, spatial distribution, and patterns of the SSA daily from day 2 through the life span of the culture on the MEA.

3. Results and Discussions

3.1 The life span of the chick FBN neuronal network in culture

3.2 The time course of the spontaneous spiking activity (SSA) of chick FBN network in culture

3.3 The spatial distribution of the SSA of chick FBN network in culture

3.4 The patterns of the chick FBN network SSA

3.1 The life span of the chick FBN neuronal network in culture

Table 1. Life span of chick FBN culture we achieved in comparison with the literature

Authors (et. al.)	Publication (Year)	Embryonic Age (Day)	Plating Density (cells/mm^2)	Coating	Medium	Research Aims	Glial Growth	Glial Inhibition	Life Span (Day)
Louis	1981	E8	700 - 800	PLL	DMEM	EM examination of synapse formation	not wanted	no	8 to 9
Tokioka	1993	E6, E8, E10, E12	~ 416	PLL	EMEM	Patch clamp study of synapse formation		no	< 14
Liu	1997	E10	~85	PLL	EMEM	Patch clamp study of a drug effect		no	7 to 30
Heidemann	2003	E8	80 - 100	PLL	M199 or L-15	low density pure forebrain neuron culture		no	< 14
We, Dr. Gao's lab	2012	E6, E7, E8	100 - 1000	PLL, or plus L1	M199	low density pure forebrain neuron culture		no	< 14
We, Dr. Gao's lab	2013	E6, E8	~ 3000	PLL+L1, or PEI	M199	high density culture for MEA experiments	wanted	no	> 29

- For low- to medium-density chick FBN culture (< 1000 cells/mm^2), the life span we can achieve is similar to other researchers' results; that is, generally no more than 14 days in vitro.
- For high-density MEA experiments (3000 cells/mm^2) without any glial inhibitory treatment, the maximum life span we have achieved so far is up to 29 days.

3.2 The time course of the spontaneous spiking activity (SSA) of chick FBN network in culture

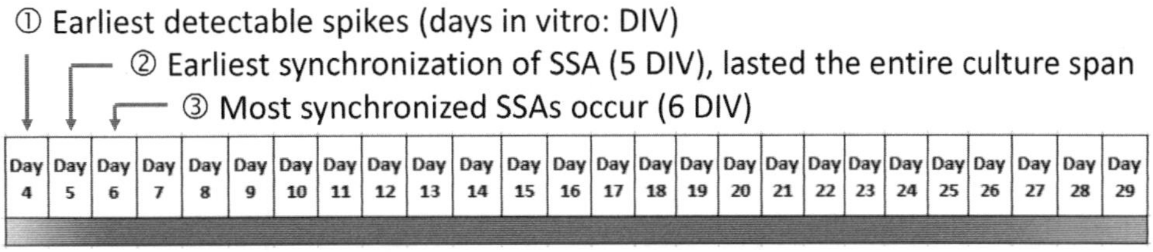

The overall strength of the activity (red — high; pink — low) is evaluated by the firing rate, the amplitude, and the spatial density of SSAs.

Fig. 1 The time course of SSA of chick FBN culture (N = 5)

3.3.a The spatial distribution of the SSA of chick FBN network in culture

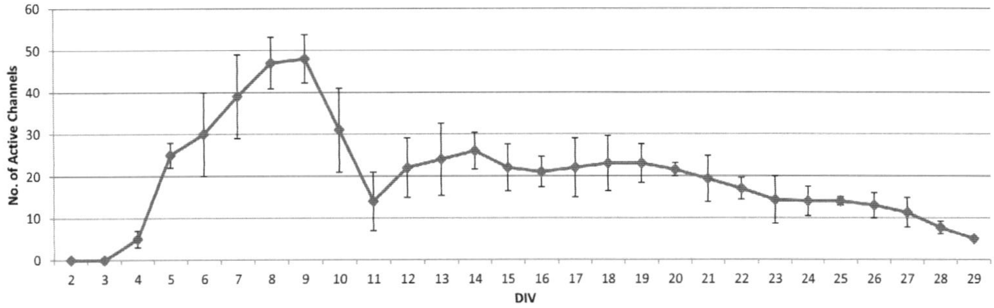

① A rapid increase in active channels to its maximum. Neurons were distributed evenly.

② Neurons aggregated to form small regional monolayers intermingled with small empty areas. Active channels underwent dramatic change in spatial distribution.

③ A delayed glial growth gradually filled out those empty areas, and an integrated cell monolayer with mixed neurons and glias formed. The spatial distribution of active channels were now relatively stable.

④ The number of active channels started to decrease.

Rapid synapse developing stage

- Network reorganization
- Solution: use E10 or E12 FBN instead of current E8 FBN to make use of the developed glial population that starts after E10.

- Network matured and SSA stabled.
- Can be used as a proper time window to conduct neurotoxin assays.

- Degenerative stage

Fig. 2 The spatial distribution of SSA of chick FBN culture (N = 6). The number of active channels along with the increase in DIV

3.3.b The spatial distribution of the SSA of chick FBN network in culture

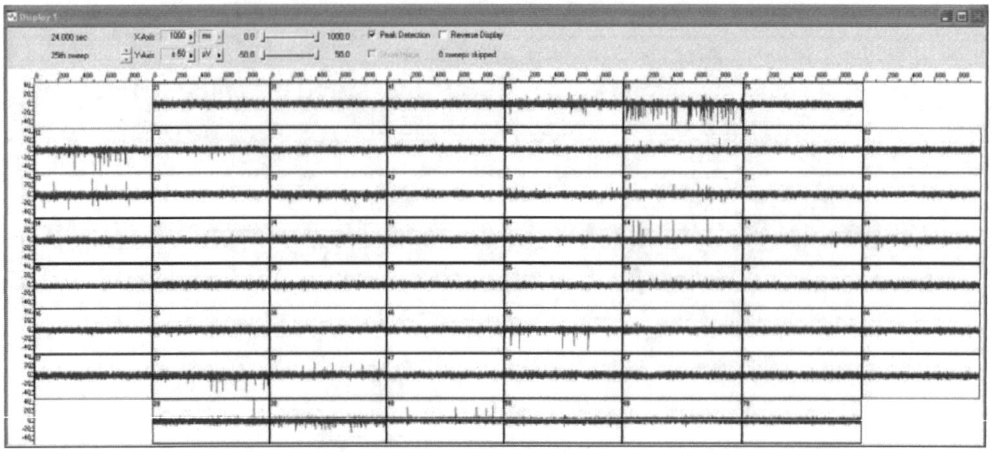

Fig. 3 **The spatial distribution of SSA of chick FBN culture:**
The distribution of active channels of one sample at 5 DIV. Unsynchronized SSAs are seen in multiple channels.

3.3.c The spatial distribution of the SSA of chick FBN network in culture

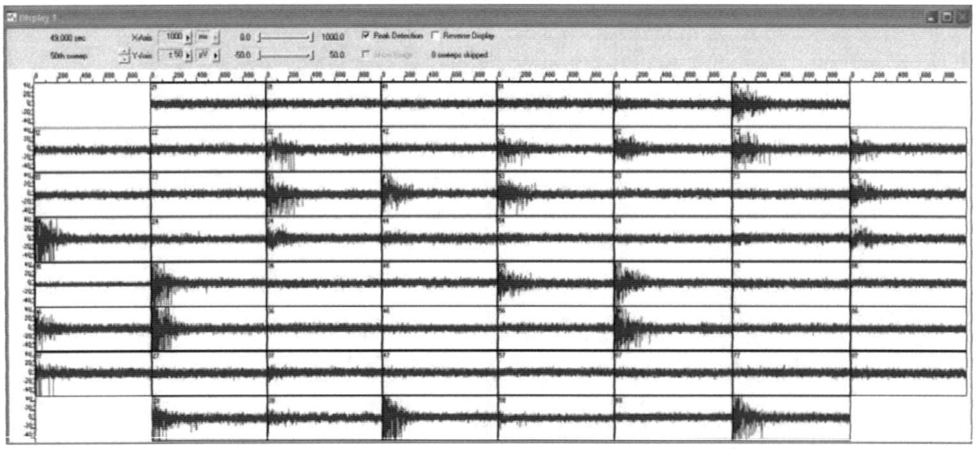

Fig. 4 **The spatial distribution of SSA of chick FBN culture:**
The distribution of active channels of one sample at 7 DIV. Synchronized SSAs are seen in multiple channels.

3.3.d The spatial distribution of the SSA of chick FBN network in culture

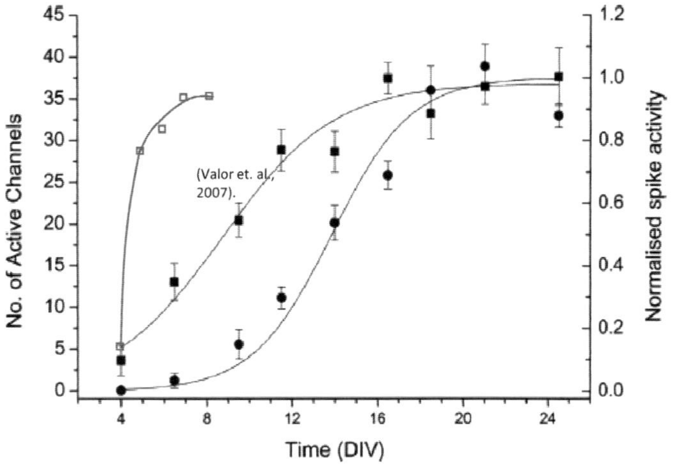

Fig. 5. The spatial distribution of SSA of chick FBN culture: A comparison with rat results. Our chick results (red) line is imposed on rat results from the Valor et al. 1) Within 8 DIV, the number of active channels of chick samples increased much faster than rats' samples. This result is consistent with Louis (1981) results, suggesting a much faster synapse formation rate in chick in vitro than in rats. 2) The maximum number of active channels of our chick samples are similar to those of rats. 3) Chick results after 8 DIV are not shown here due to its unstable network reorganization described previously in Fig. 2(A). We expect E10 or even E12 chick samples to generate better results in future.

3.4 Conclusion on the patterns of the chick FBN network SSA

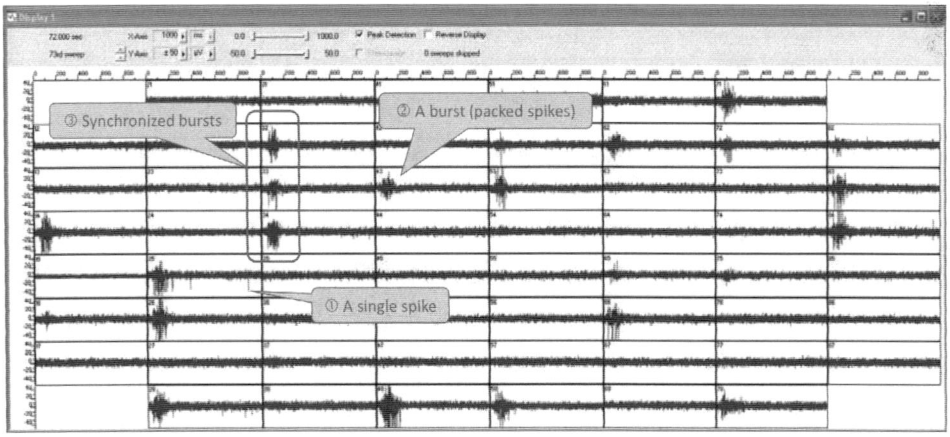

Fig. 6 The patterns of SSA of chick FBN network: Patterns exhibition similar signals as seen in rats and mice, for example, the firing of SSA is always paroxysmal.

Acknowledgments

This research was supported by the National Basic Research Program of China (973program, 2009CB930004), the National Natural Science Foundation of China (31070847), and the Guangdong Provincial Department of Science and Technology of China (2011B050400011).

Reprogramming Life: The Emergence of Synthetic Biology

James J. Collins

Howard Hughes Medical Institute
Dept of Biomedical Engineering & Center for BioDynamics
Boston University

Wyss Institute for Biologically Inspired Engineering
Harvard University

Synthetic Biology: Engineered Gene Networks

Design & model network | **Encode into DNA plasmid** | **Transfer to cell** | **Test network dynamics**

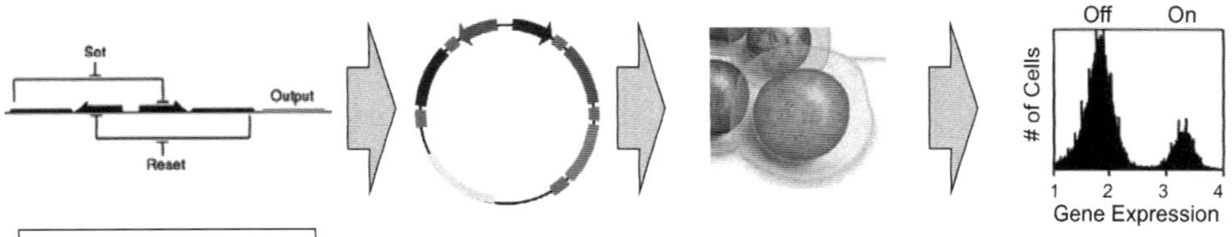

$$\frac{du}{d\tau} = \frac{\alpha_1}{1+v^{\beta}} - u$$

$$\frac{dv}{d\tau} = \frac{\alpha_2}{1+u^{\gamma}} - v$$

TS Gardner et al., Nature, 2000

Schematic Design of Genetic Toggle Switch

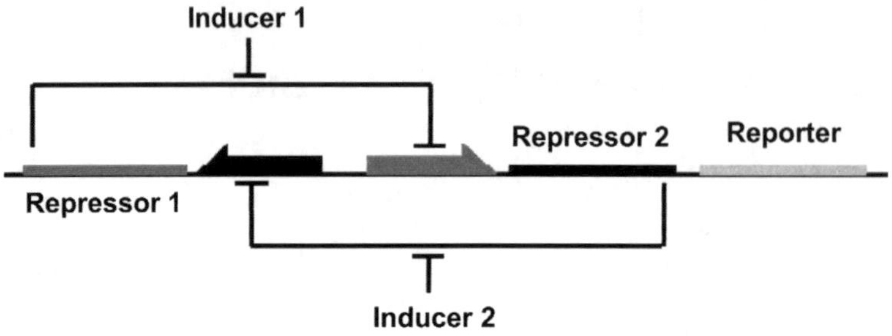

TS Gardner et al., Nature, 2000

BUEngineering

Genetic Toggle Switch: Results

$$\dot{u} = \frac{\alpha_1}{1 + v^{\beta}} - u$$

$$\dot{v} = \frac{\alpha_2}{1 + u^{\gamma}} - v$$

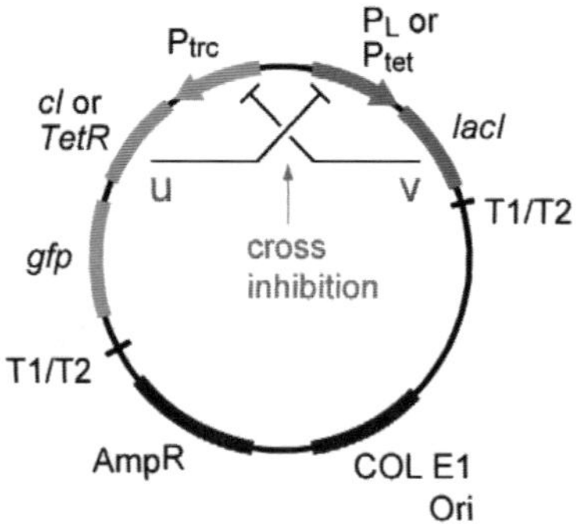

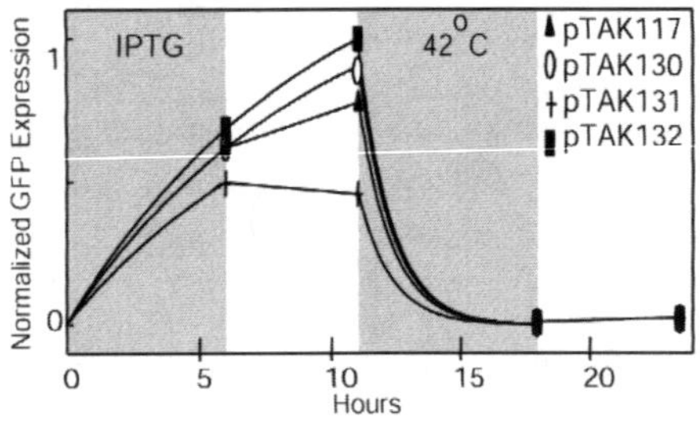

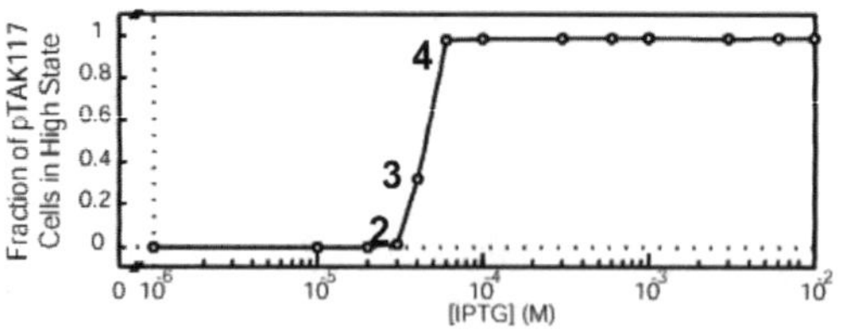

BUEngineering

Programmable Cells

Interfacing natural and engineered gene networks

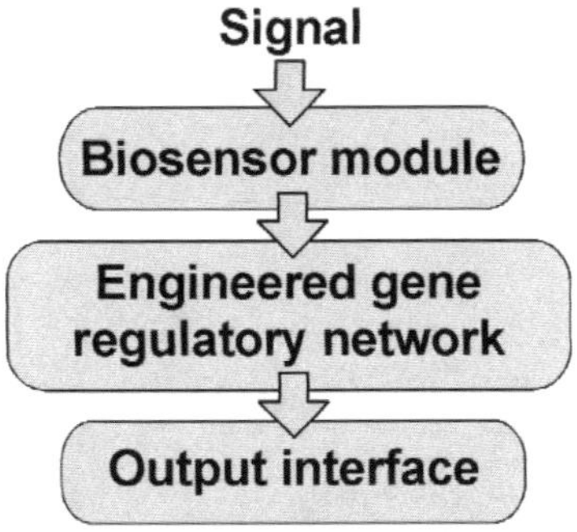

Module function

Sensor: converts biological signal *S* into input $I(S)$

Regulator: converts $I(S)$ into output $O(S)$ in accordance with its engineered properties

Output: converts $O(S)$ into biological response

H Kobayashi et al., PNAS, 2004

BU College of Engineering

Programmable Cells: Whole-Cell Biosensors

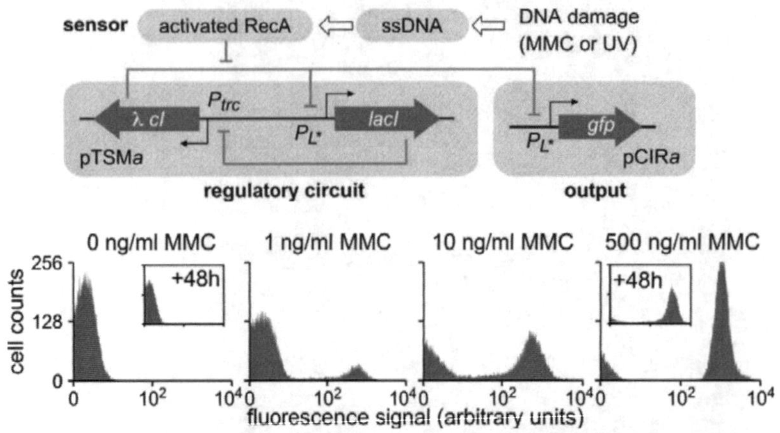

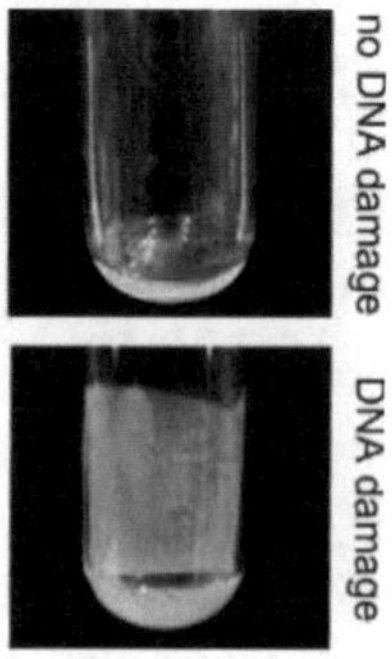

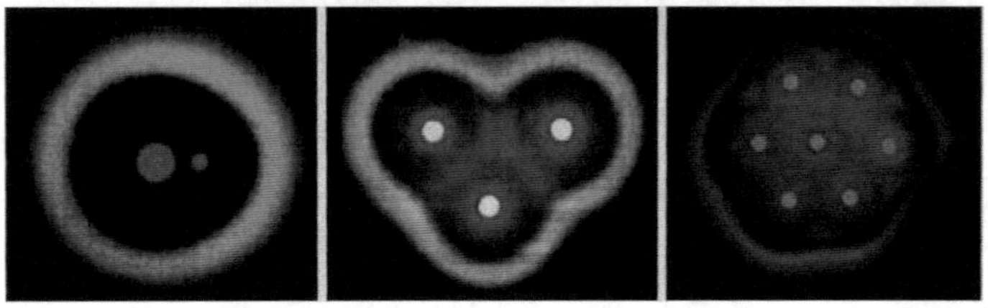

From Ron Weiss

BUEngineering

Programmable Cells: Crowd Sensor

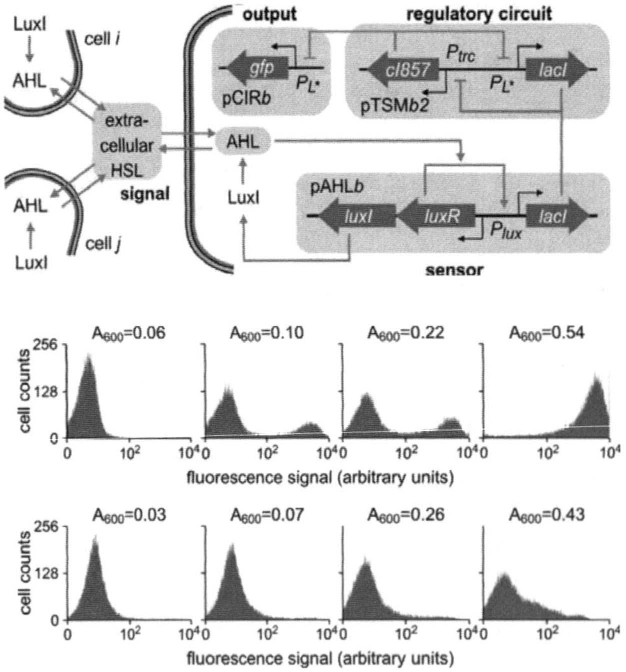

BUEngineering

RNA-based Synthetic Biology

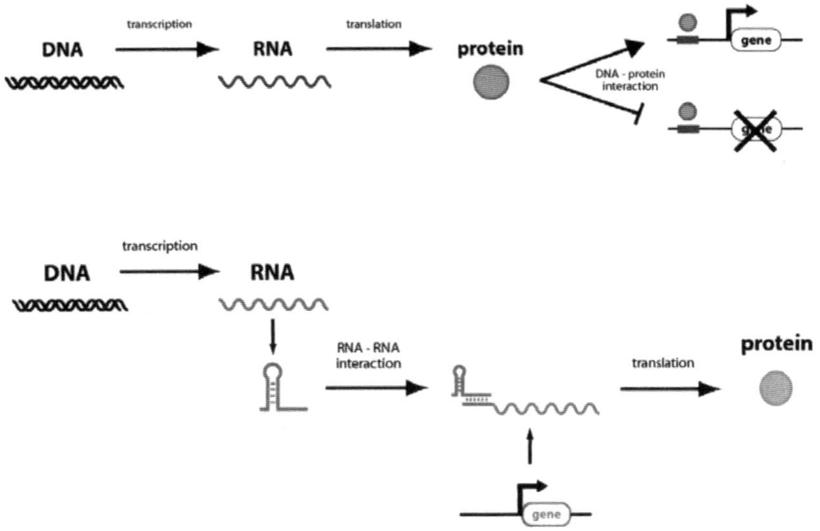

FJ Isaacs et al., Nature Biotechnology, 2006

BUEngineering

RNA Switches: Engineered Riboregulators

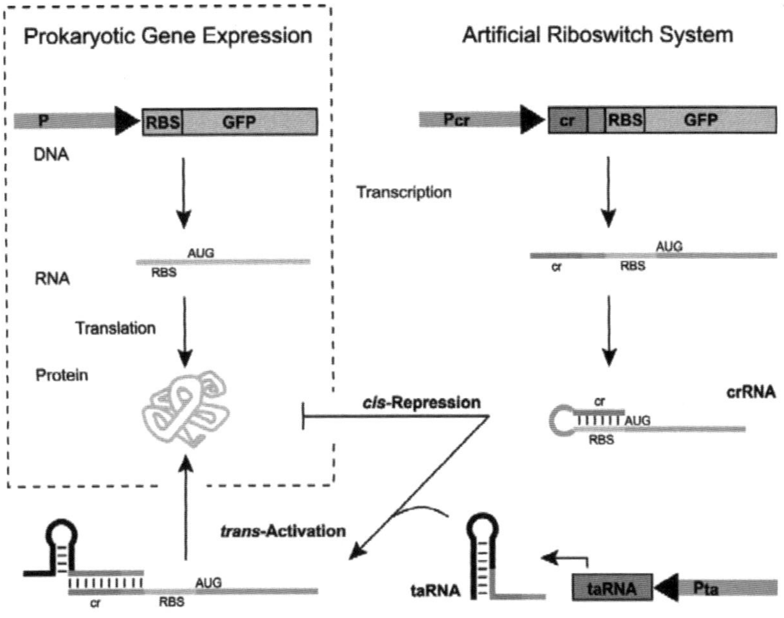

FJ Isaacs et al., Nature Biotechnology, 2004

BUEngineering

Engineered Riboregulator: Trans-Activation

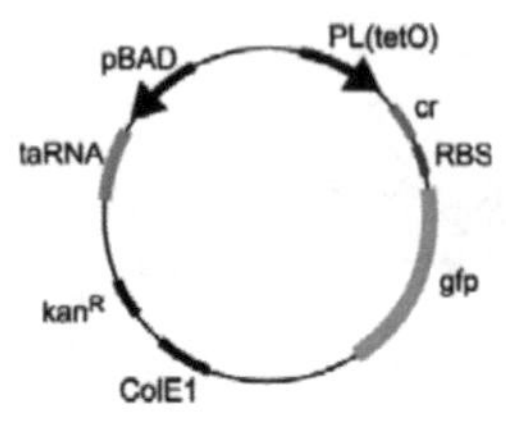

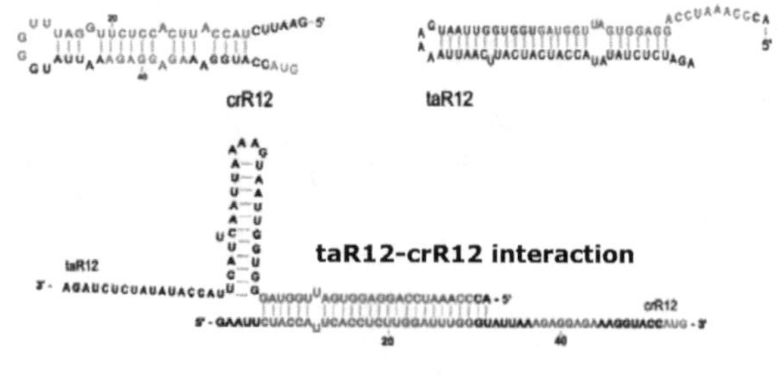

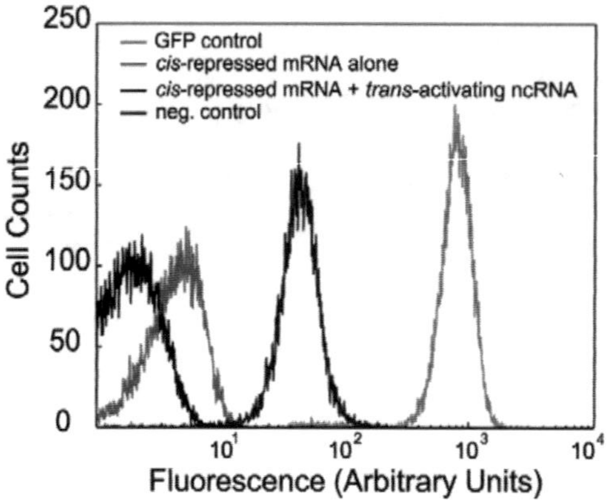

BUEngineering

Engineered Riboregulator: System Performance

Transient Response

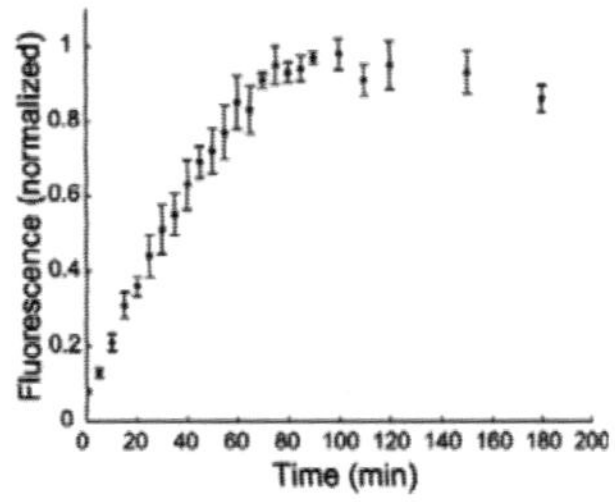

Steady-State Response

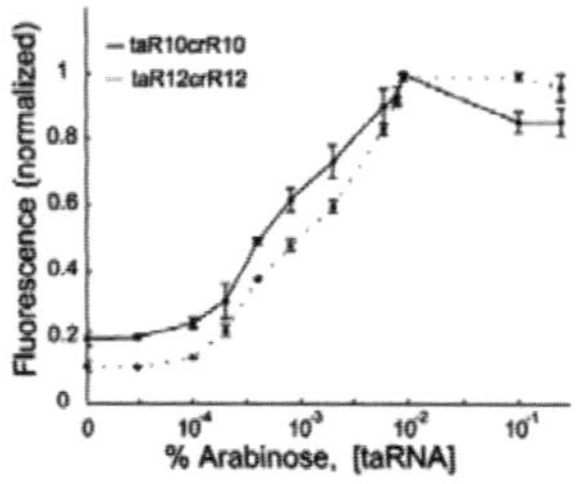

Specificity

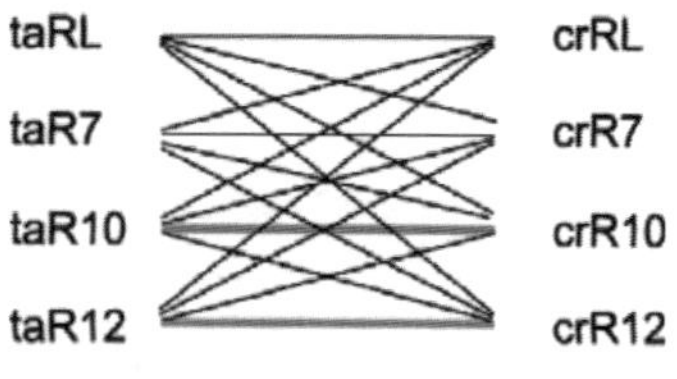

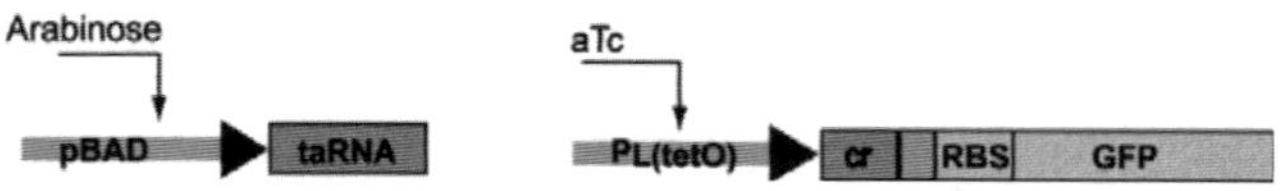

FJ Isaacs et al., Nature Biotechnology, 2004

BUEngineering

Genetic Switchboards: Dynamically Tuning Cellular Metabolism

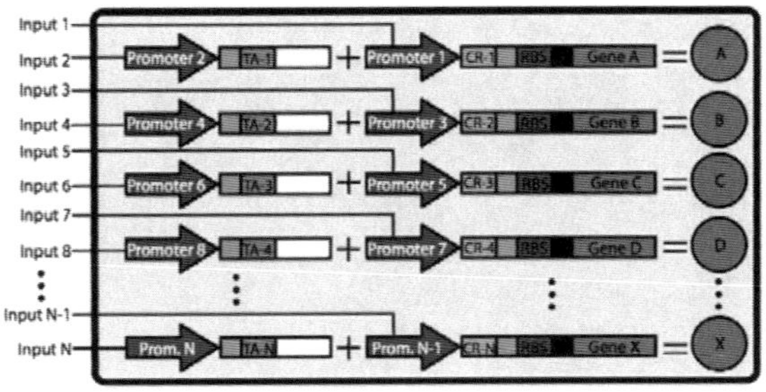

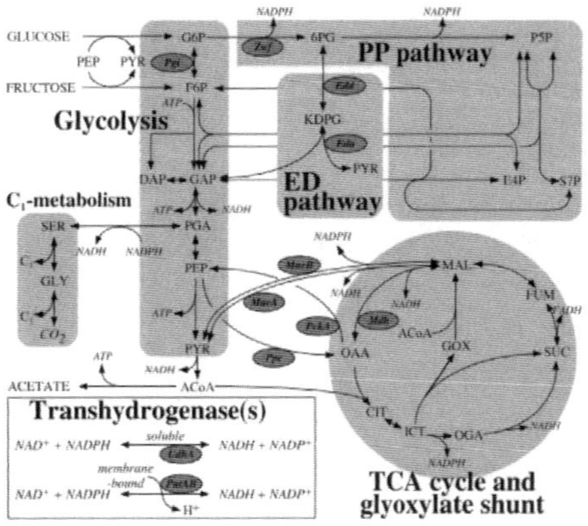

Programmable microbes for bioreactors: bioenergy & biomaterials

JM Callura et al., PNAS, 2012

Synthetic Biology: Ethical and Social Concerns

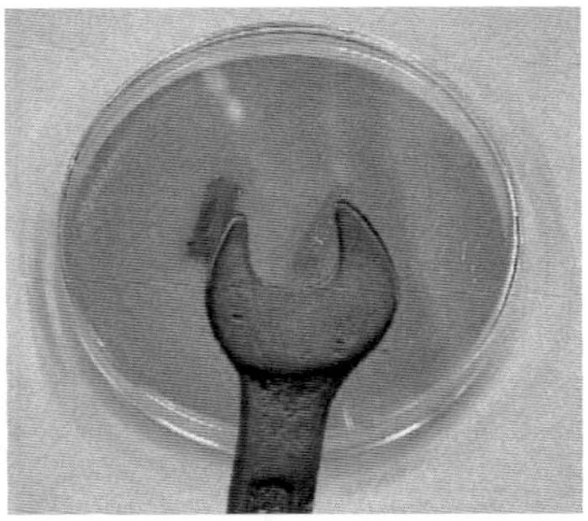

Synthetic Gene Networks that Count: RNA Switches

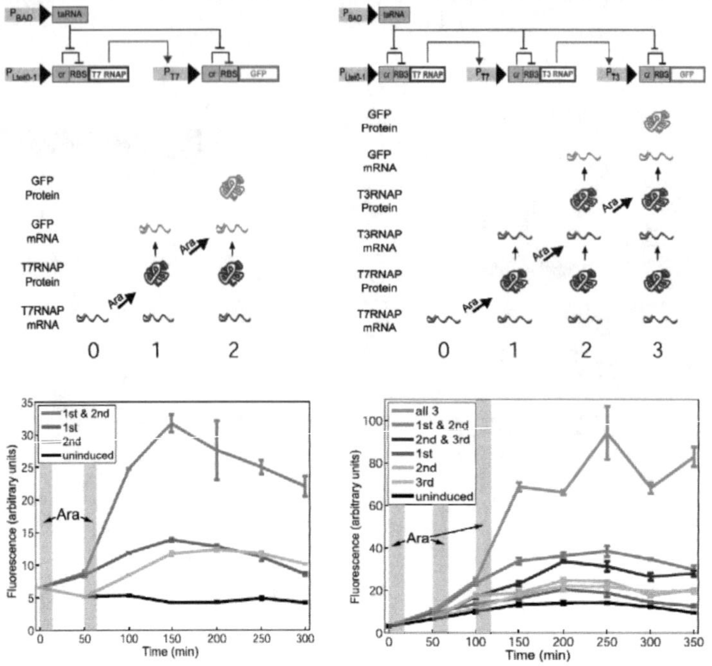

AE Friedland et al., Science, 2009

BUEngineering

Genetic Counter: Recombinase-Based Cascade of Memory Elements

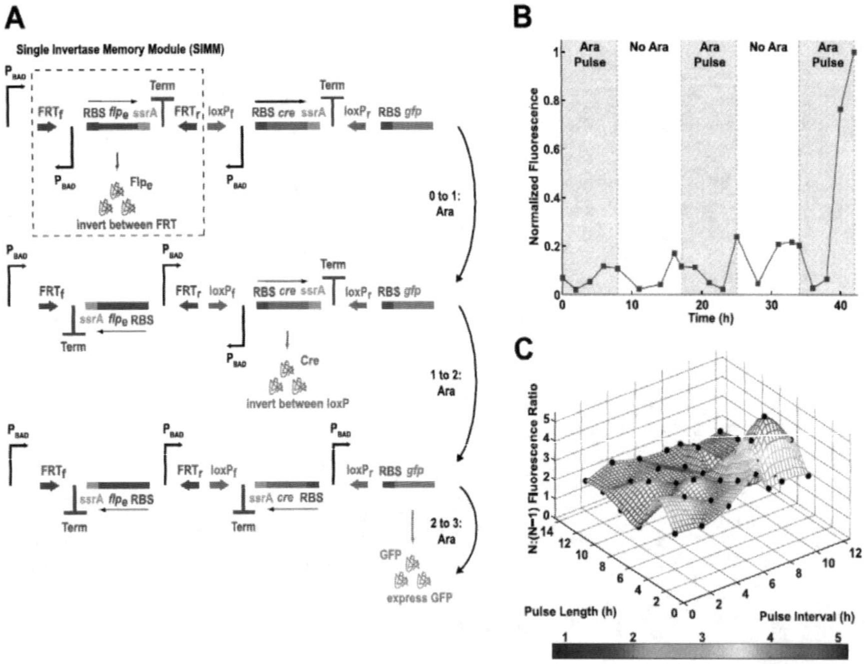

AE Friedland, TK Lu et al., Science, 2009

BUEngineering

Programmable Kill Switch for Microbes

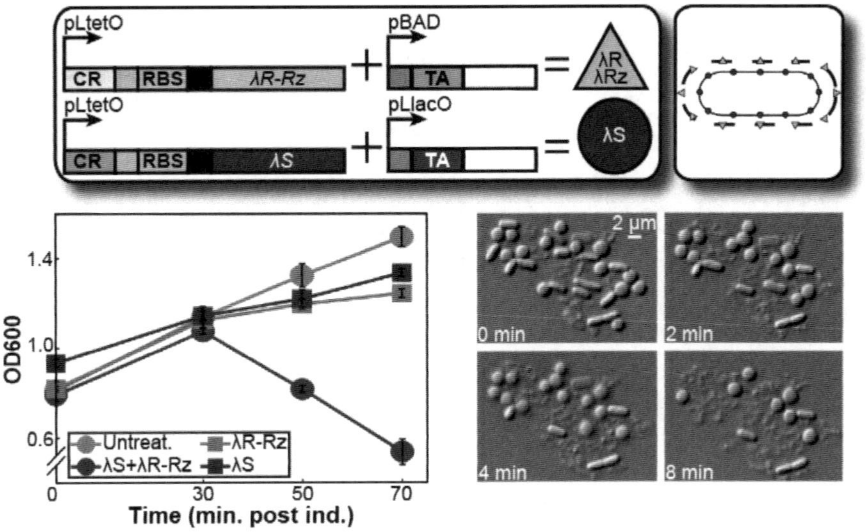

JM Callura et al., PNAS, 2010

Synthetic Gene Networks: Significant Bottleneck

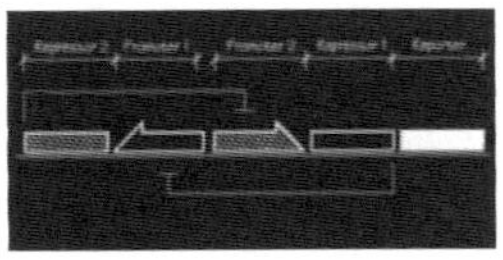

Gene circuit blueprint

$$\frac{du}{d\tau} = \frac{\alpha_1}{1 + v^{\beta}} - u$$

$$\frac{dv}{d\tau} = \frac{\alpha_2}{1 + u^{\gamma}} - v$$

In silico model

Plasmid design

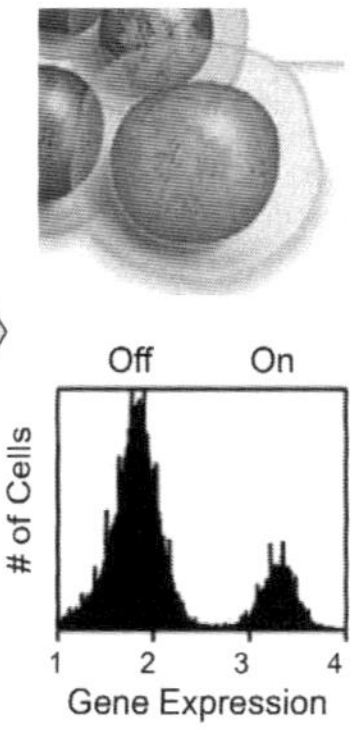

Component Libraries: Synthesis, Screening and Characterization

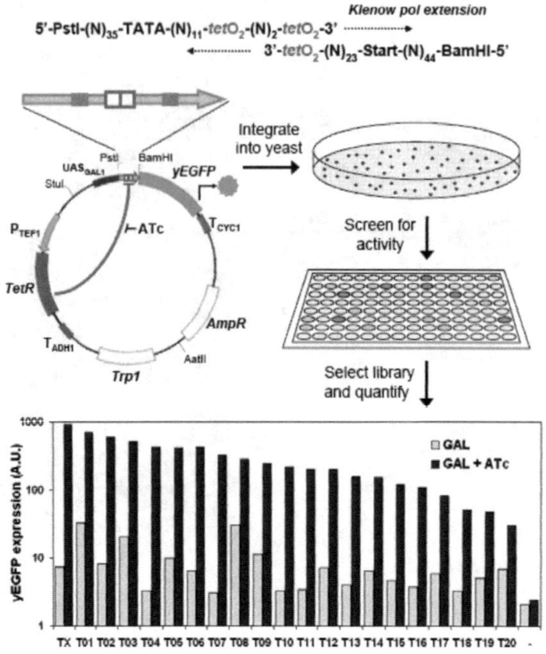

BUEngineering

Synthetic Gene Networks with Predicted Functions

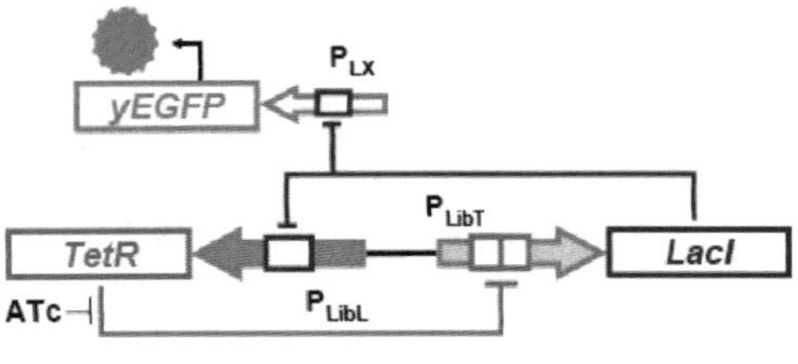

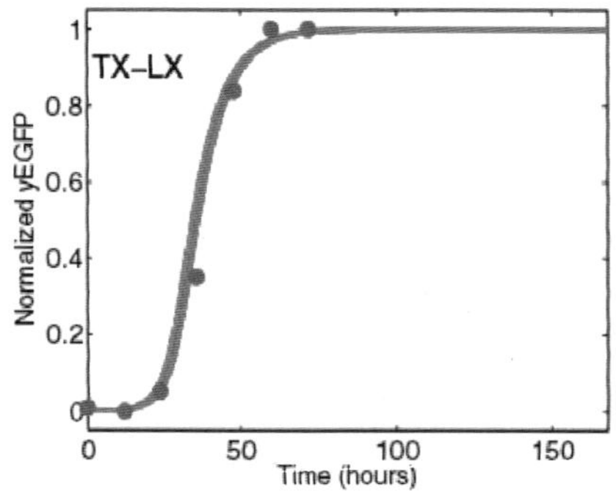

Build One, Fit One,
Predict All

T Ellis et al., Nature Biotechnology, 2009

BUEngineering

Genetic Timers with Predicted Functions

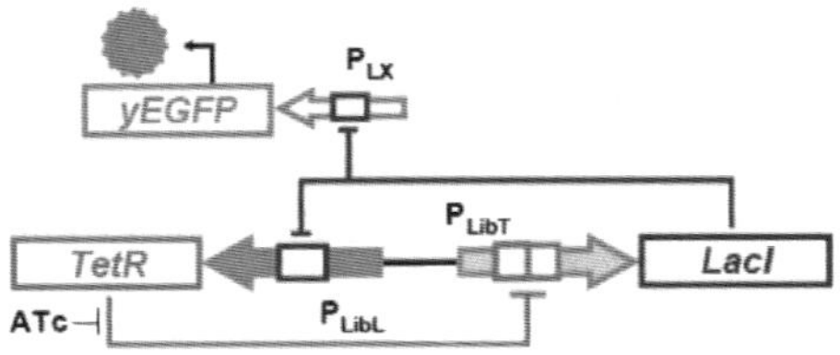

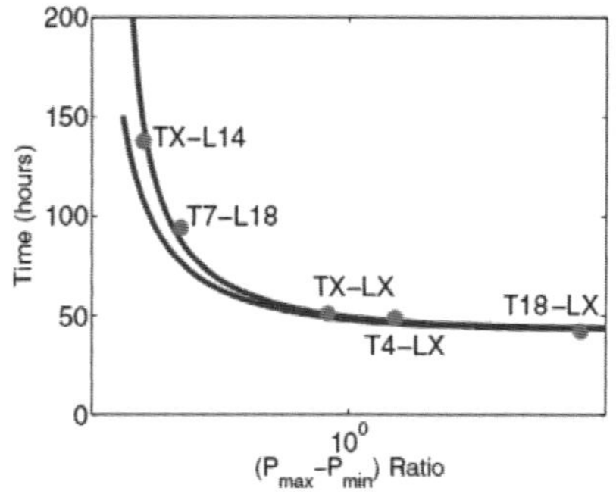

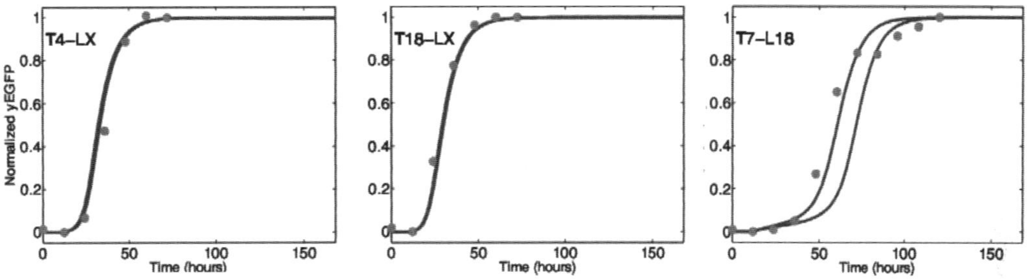

BUEngineering

Control of Yeast Sedimentation Timing

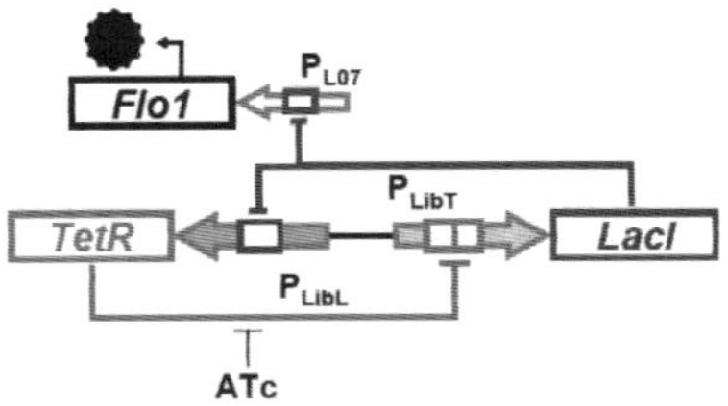

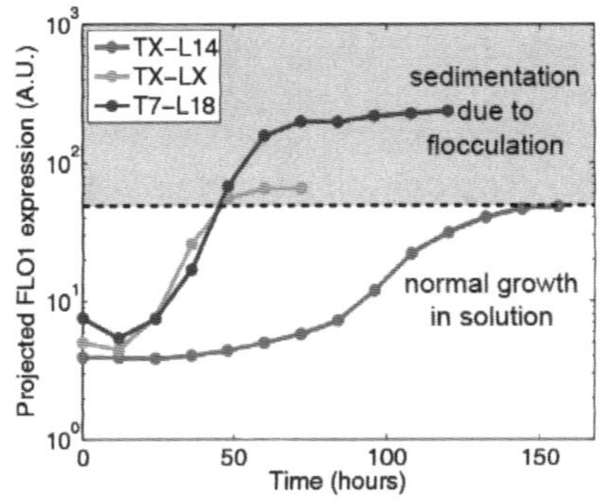

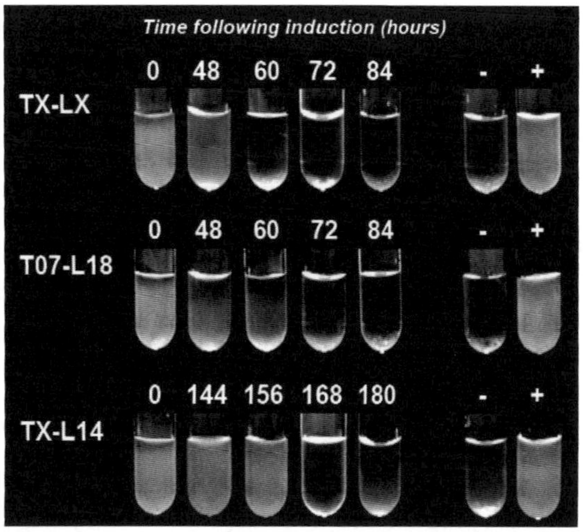

T Ellis et al., Nature Biotechnology, 2009

BUEngineering

Engineering Better Beer

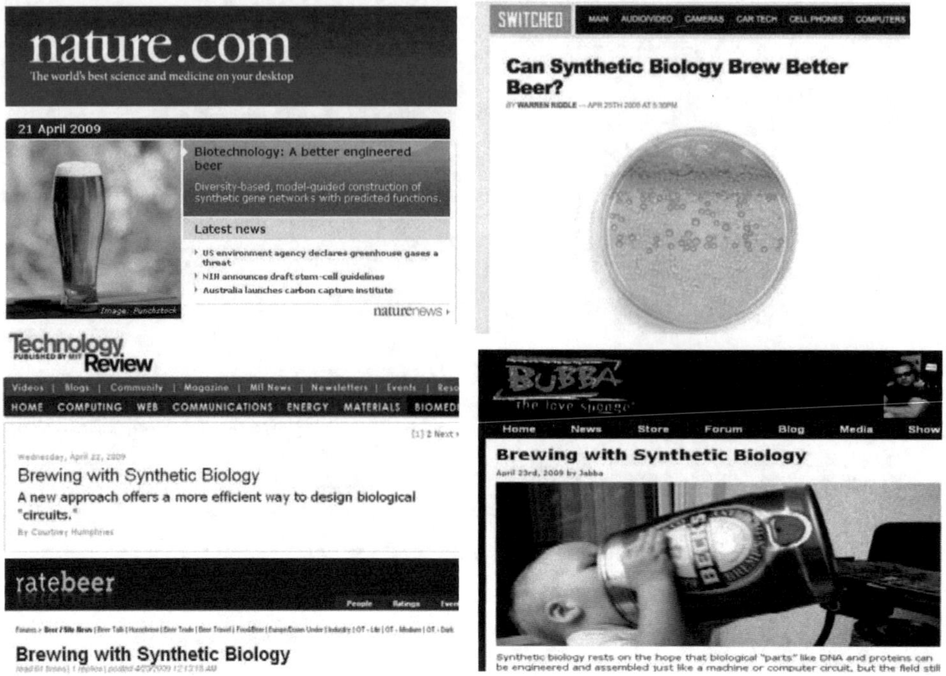

Therapeutic Synthetic Biology

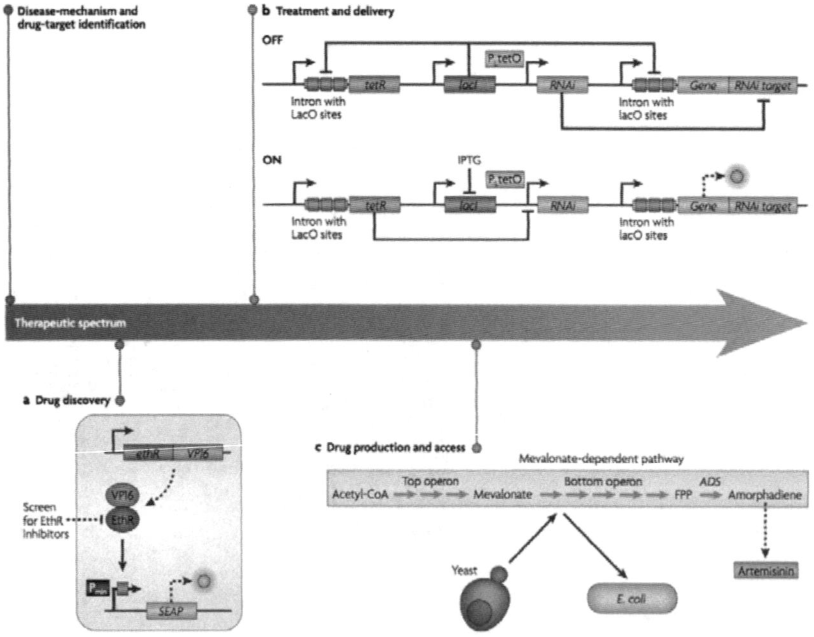

AS Khalil and JJ Collins, Nature Reviews Genetics, 2010

Synthetic Biology Meets Systems Biology

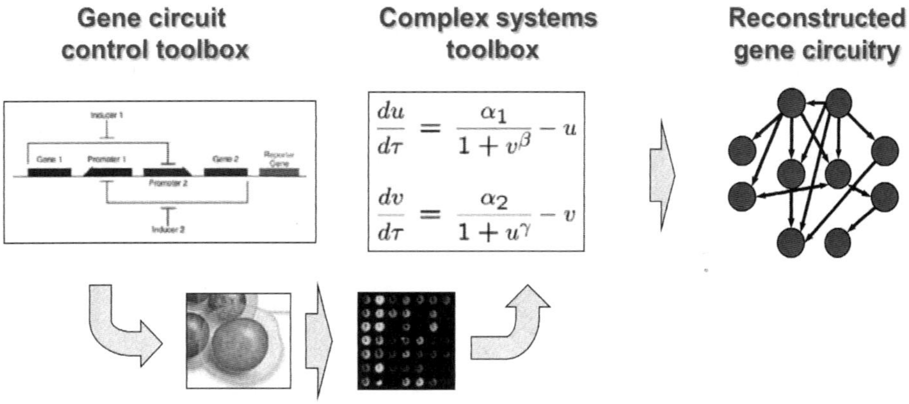

MKS Yeung et al., PNAS, 2002
J Tegner et al., PNAS, 2003

BUEngineering

Reverse-Engineered Networks and Drug-Target Identification

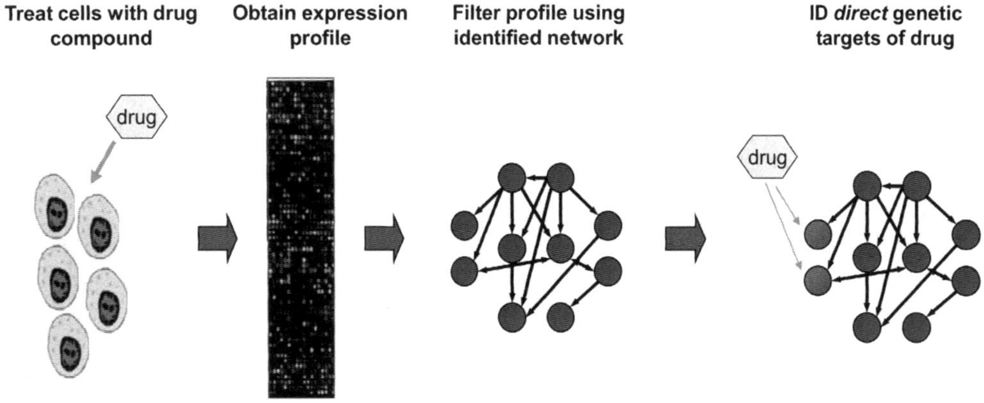

T Gardner et al., Science, 2003
D di Bernardo et al., Nature Biotechnology, 2005

BUEngineering

More Antibiotic Resistance, Fewer New Antibiotics

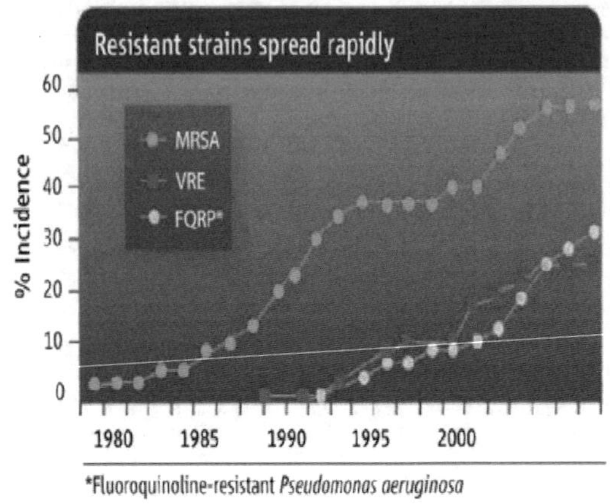

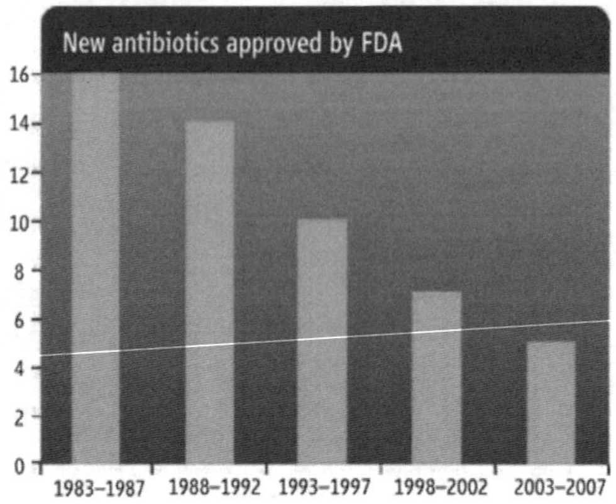

Taubes, Science, 2008

BUEngineering

E. Coli Network Reconstruction on a Genome Scale

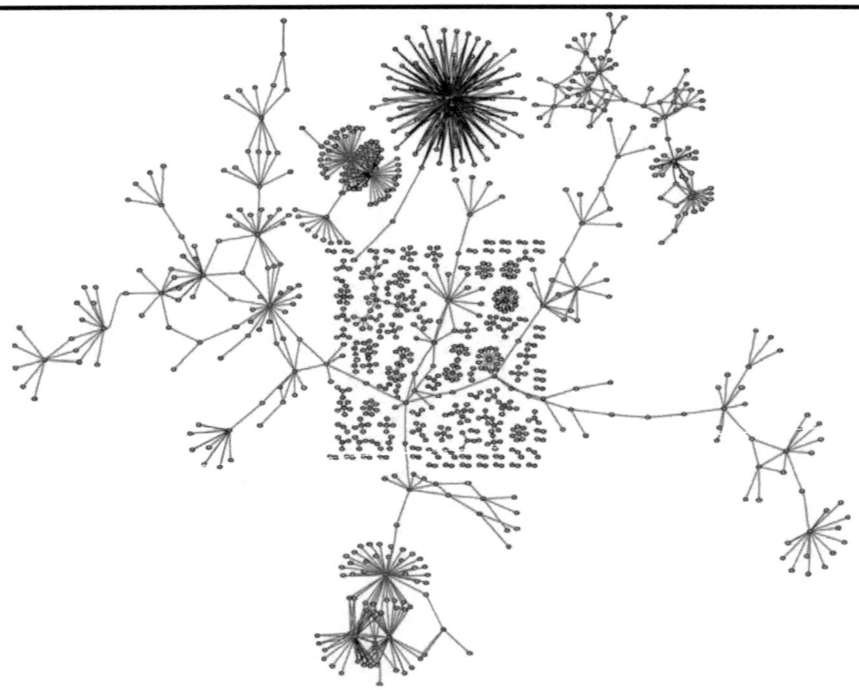

J Faith et al., PLoS Biology, 2007

BUEngineering

Synthetic Biology Study of Gyrase Inhibitors

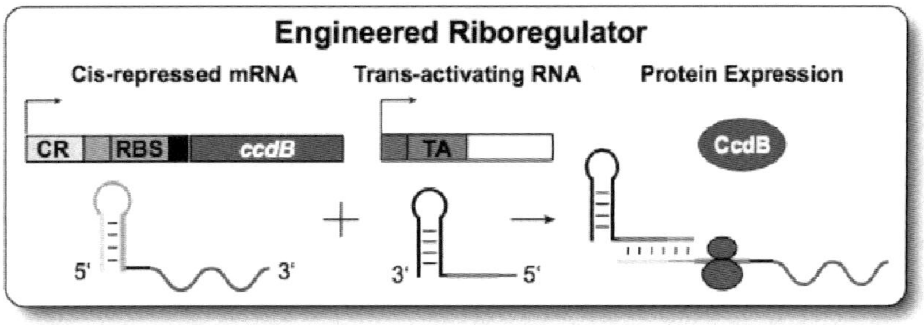

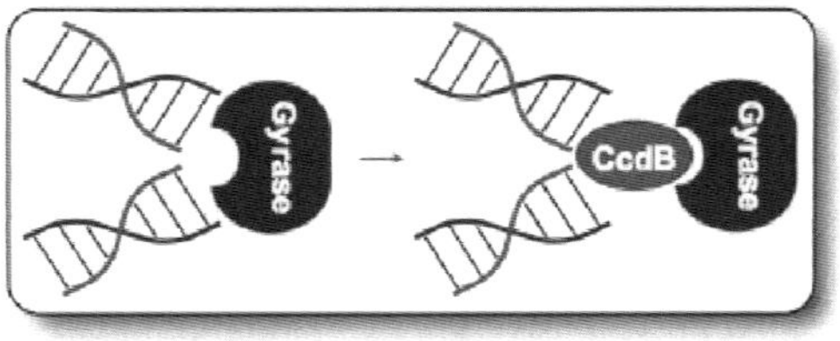

BUEngineering College of

Systems Biology and Synthetic Biology: Antibiotic Action

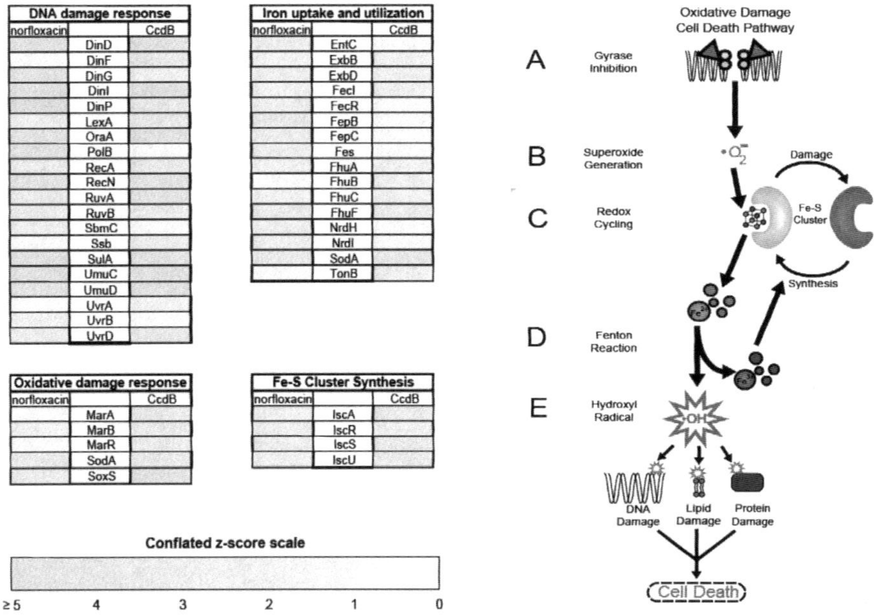

DJ Dwyer et al., Molecular Systems Biology, 2007

BUEngineering College of

A Common Mechanism Induced by Bactericidal Antibiotics

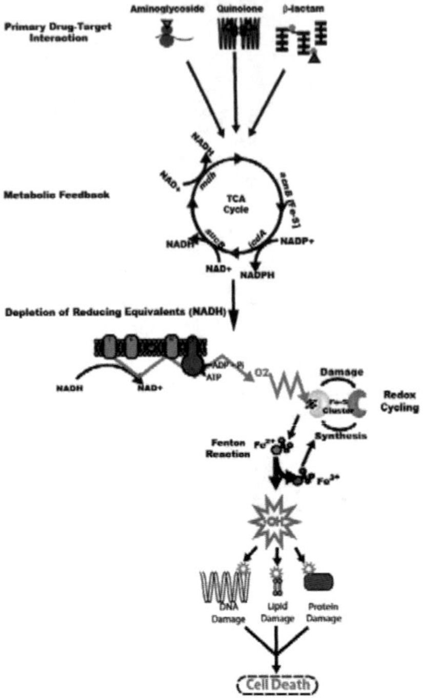

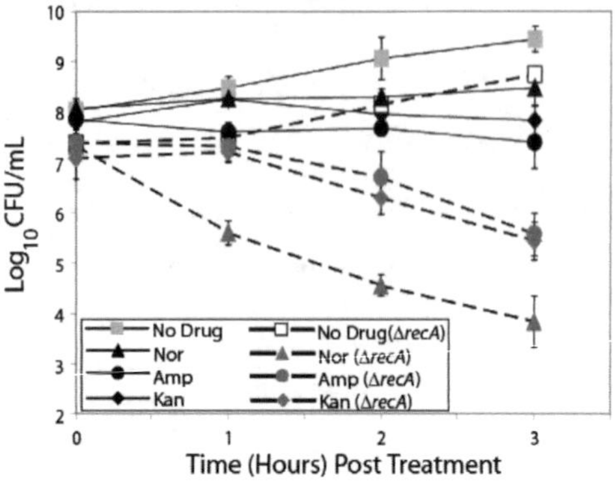

MA Kohanski et al., Cell, 2007

BUEngineering College of

RecA ATPase Inhibition: A High-Throughput Assay

BU Ctr for Chemical Methodology & Library Development:

>2,000 bioactive compounds

New England Regional Center of Excellence in Biodefense and Emerging Infectious Disease (HMS):

>50,000 bioactive compounds

Screening results:

several hits

validated inhibitors:
CB301
CB501

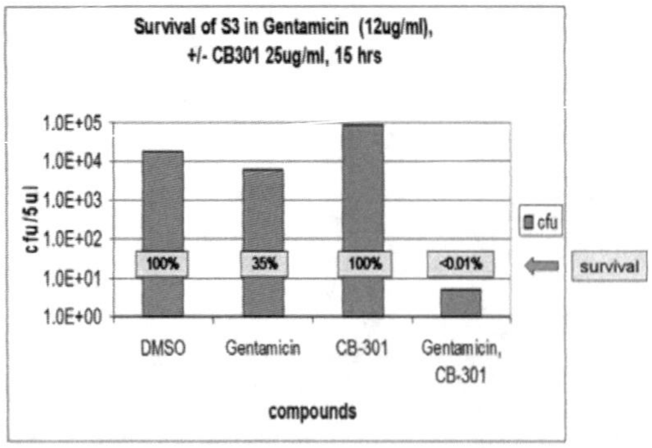

BUEngineering College of

Engineered Bacteriophage as Antibacterial Adjuvants

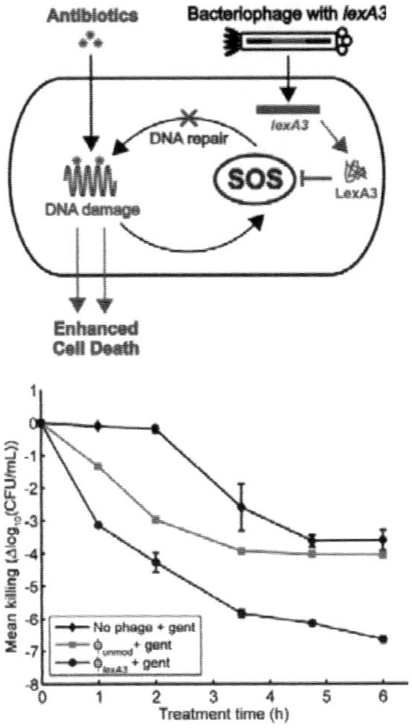

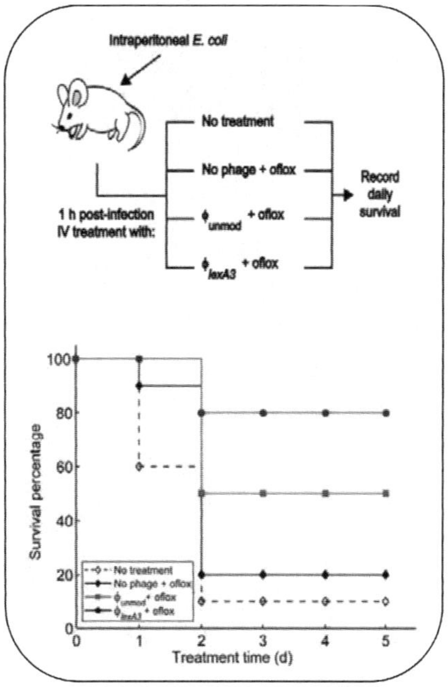

TK Lu and JJ Collins, PNAS, 2009

BUEngineering

Enter and Destroy the Biofilm Matrix

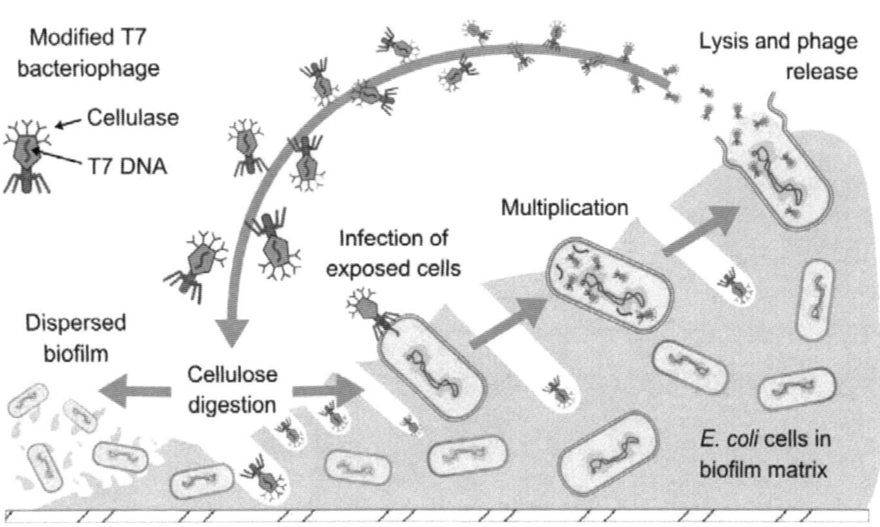

BUEngineering

Results: Engineered Enzymatically Active Bacteriophage

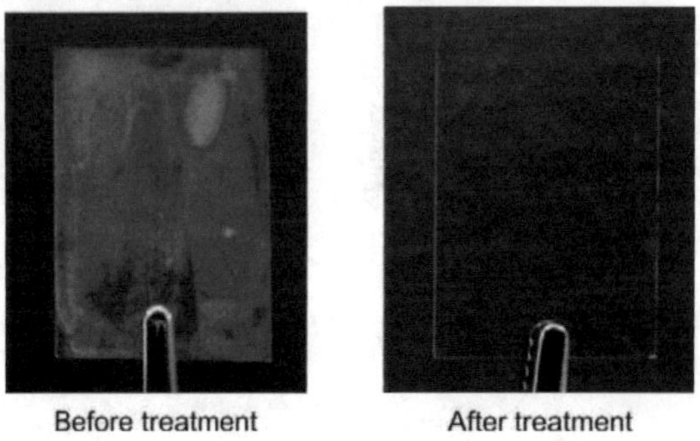

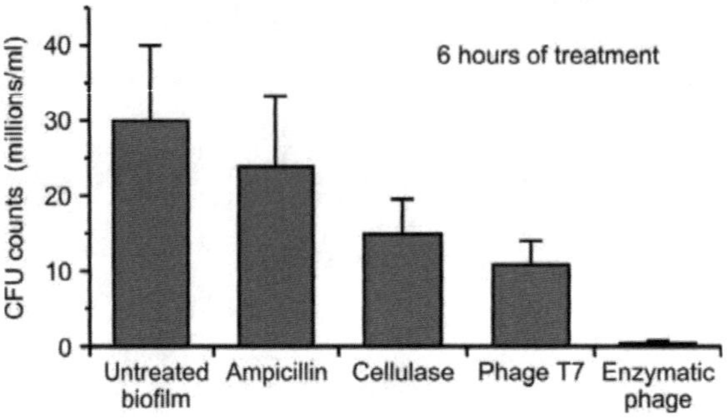

BUEngineering

Engineered Enzymatic Bacteriophage: Biofilm Busters

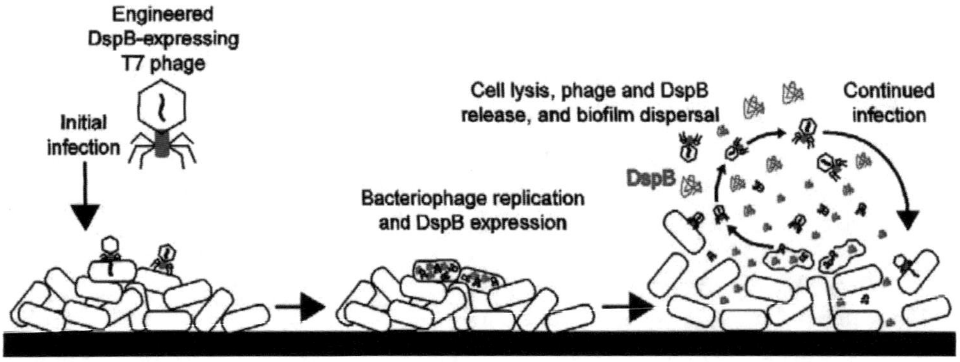

Lu and Collins, PNAS, 2007

BUEngineering

Novel Antimicrobials Through Microbial Warfare

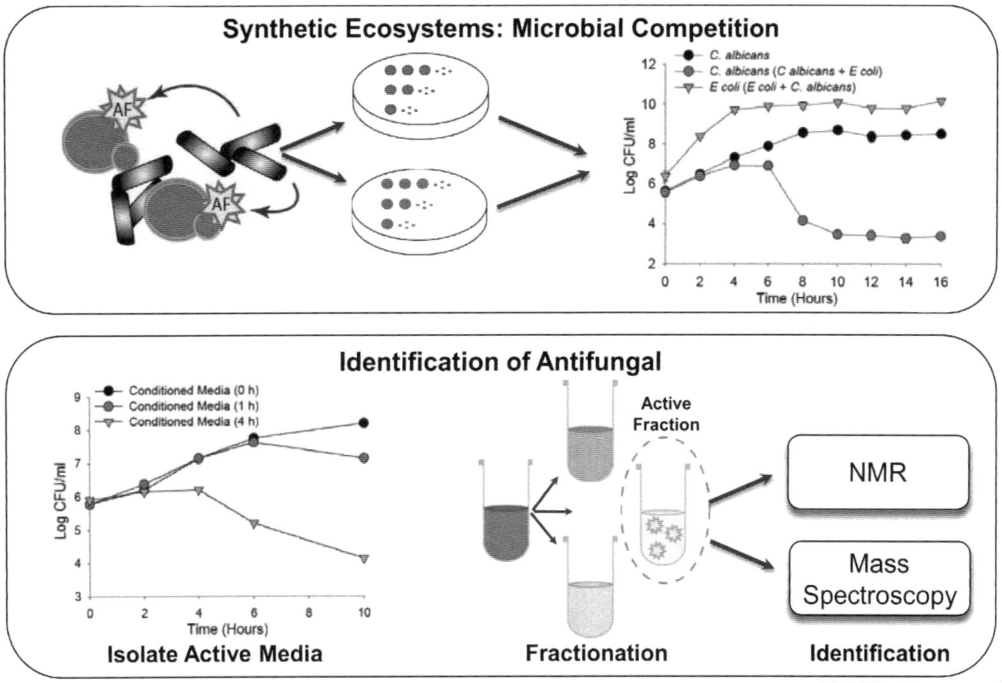

BUEngineering

Synthetic Gene Networks: Moving Up the Food Chain

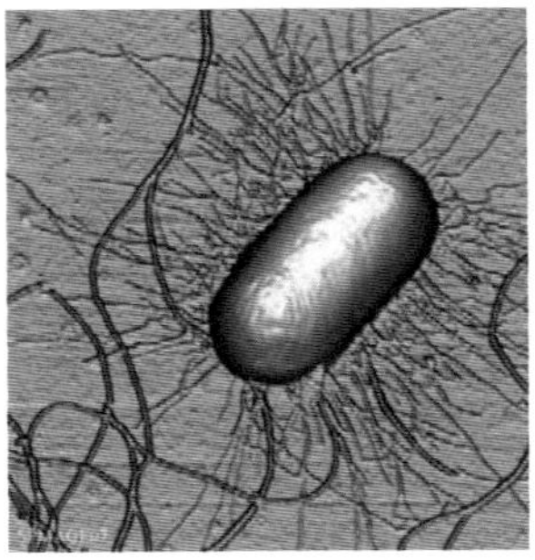

Bacteria

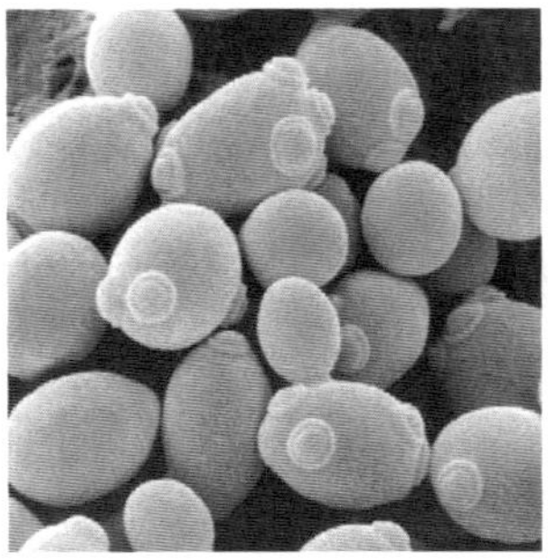

Yeast

Higher Organisms

BUEngineering

Engineered Mammalian Gene Switch: RNAi and Repressor Proteins

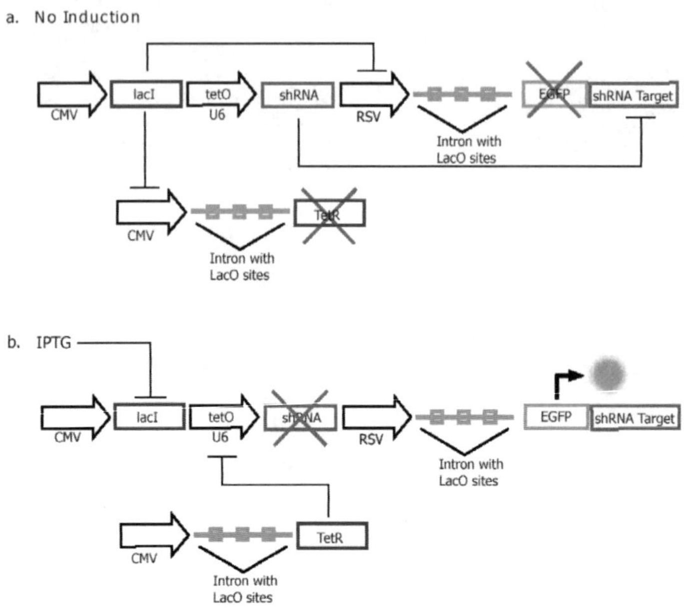

TL Deans et al., Cell, 2007

Engineered Mammalian Gene Switch: Performance Characteristics

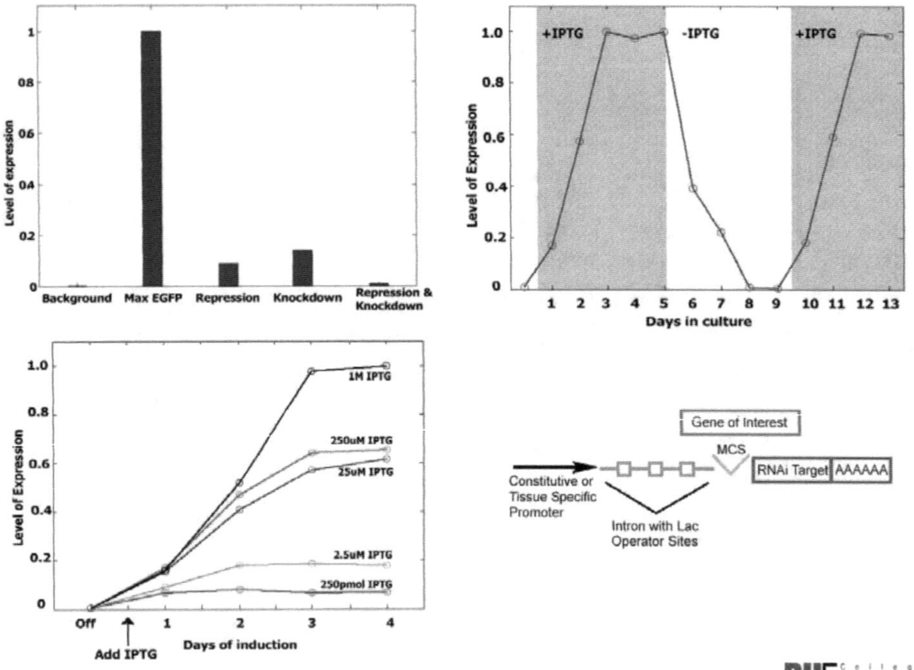

Using the Genetic Switch to Control the Expression of DTA

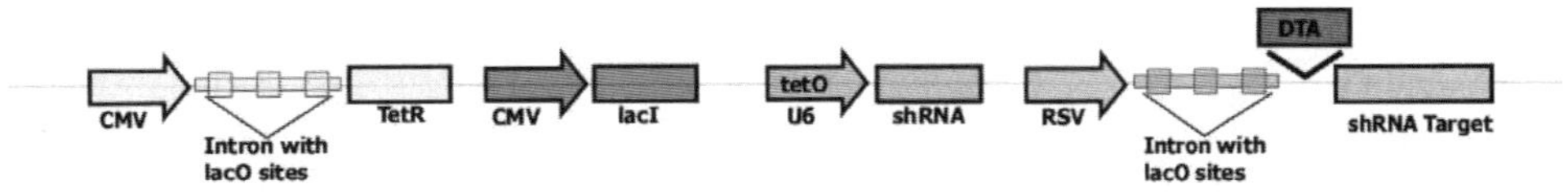

Switch Off (DTA off)

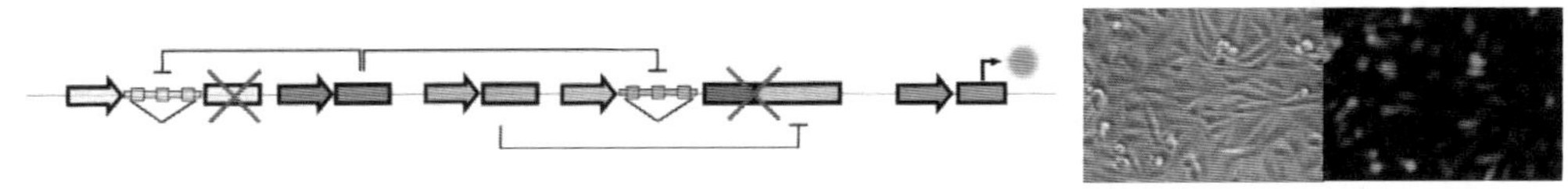

Switch On (DTA on)

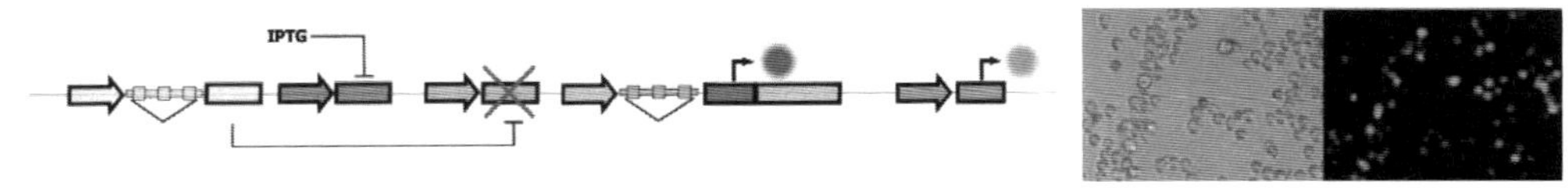

BUEngineering

Using the Genetic Switch to Control the Expression of Bax

a. Replacing EGFP with Bax

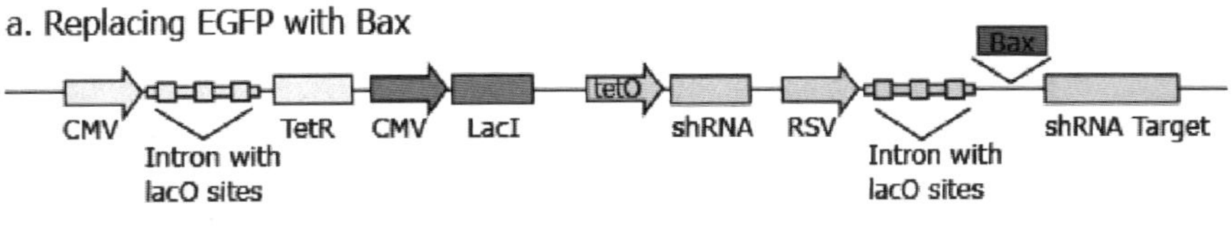

b.

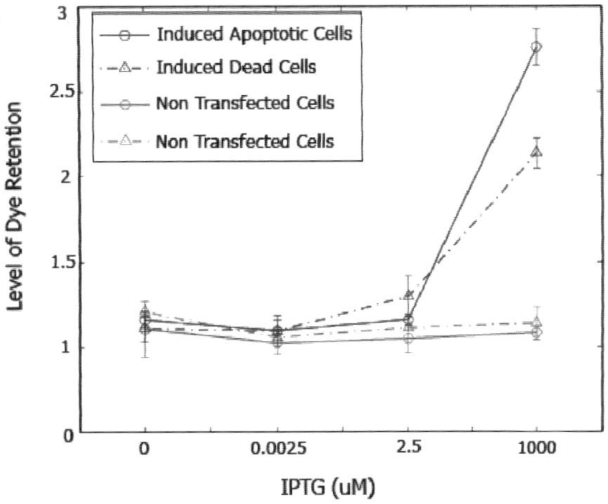

BUEngineering

Reprogramming Somatic Cells into Pluripotent Stem Cells (iPSCs)

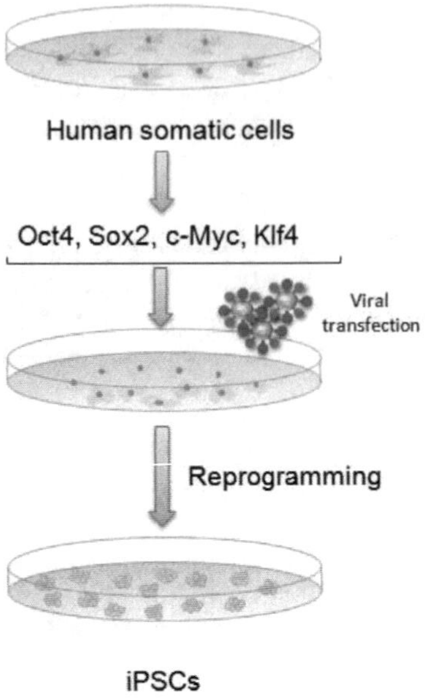

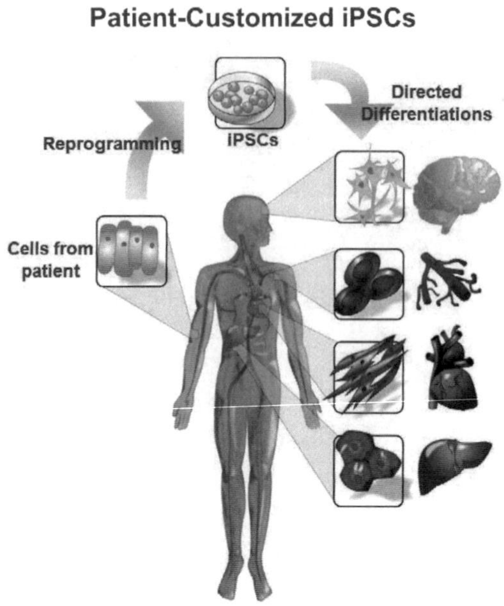

BUEngineering

Synthetic mRNA Reprogramming: RiPS

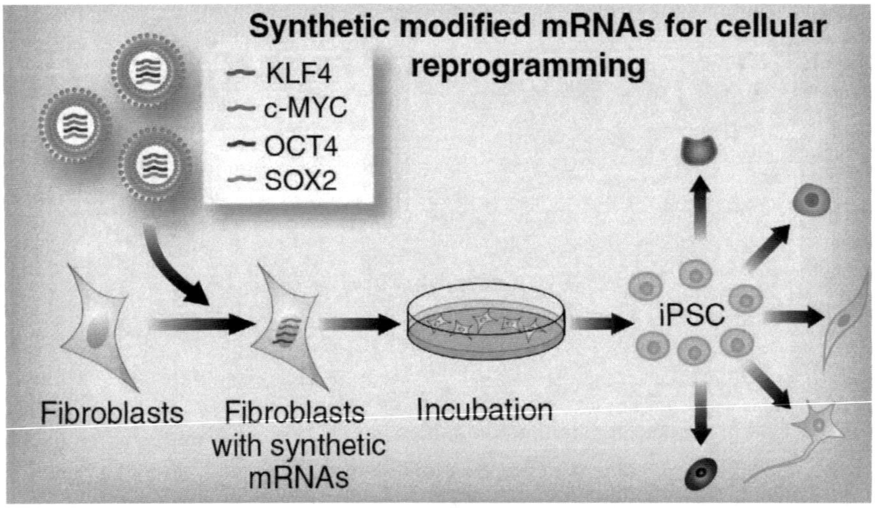

mRNA Reprogramming Kit now available from Stemgent

L Warren et al., Cell Stem Cell, 2010

BUEngineering

Creating iPSCs from Human Peripheral Blood Cells

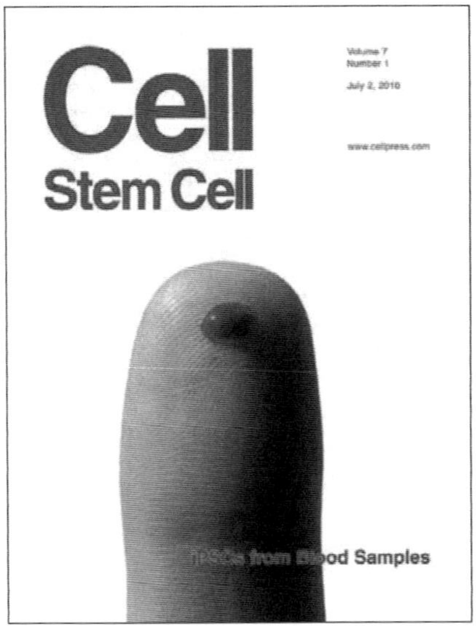

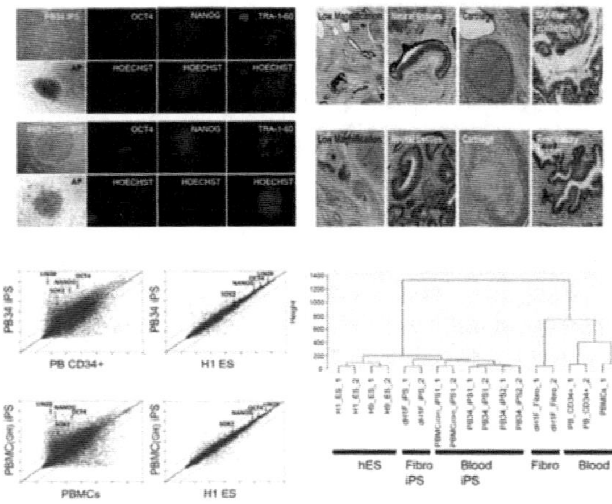

YH Ho et al., Cell Stem Cell, 2010

BUEngineering

Multistability and Small Gene Regulatory Networks

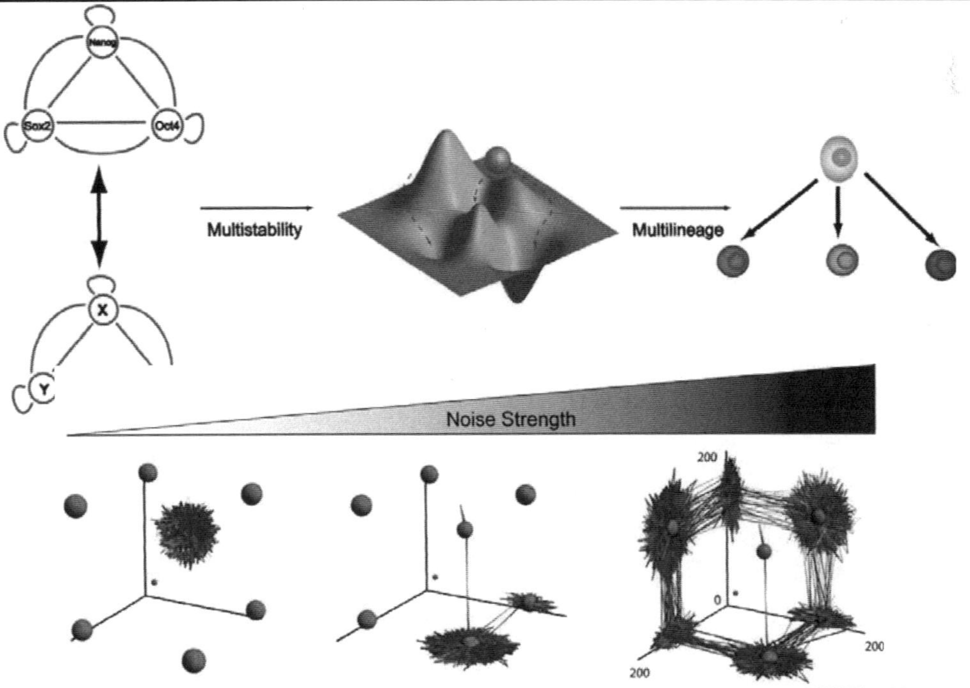

BUEngineering

Synthetic Gene Network: Tunable Noise Generator

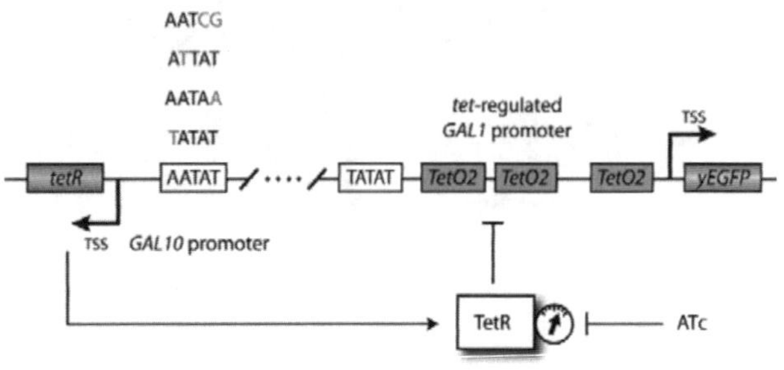

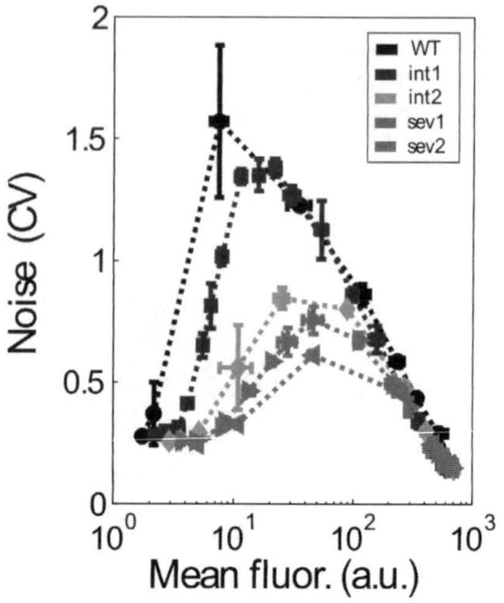

KF Murphy et al., Nucleic Acids Research, 2010

BU College of Engineering

Re-engineered Microbiomes

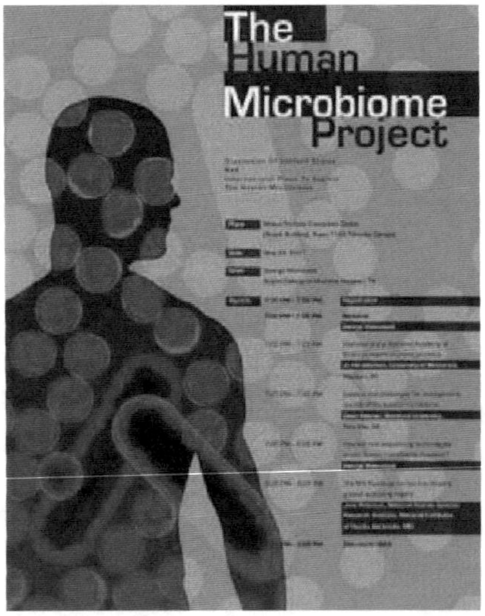

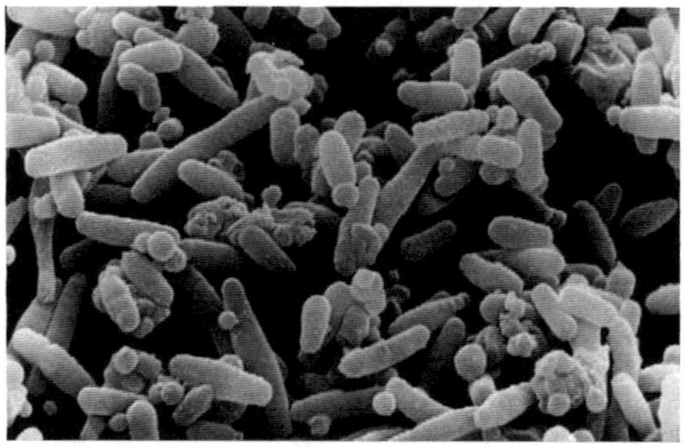

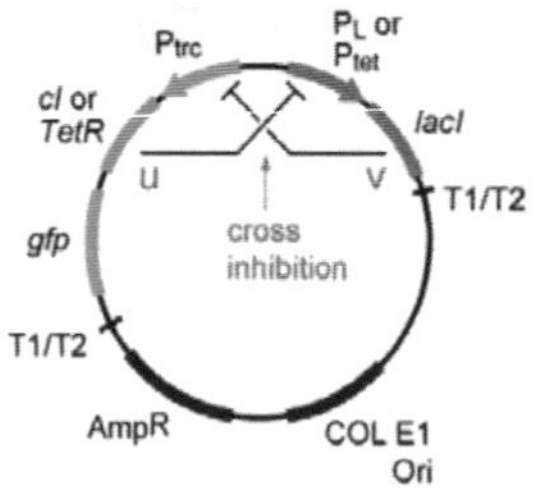

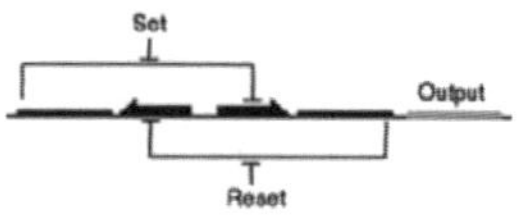

BU College of Engineering

Synthetic Probiotic for Identifying and Preventing Cholera

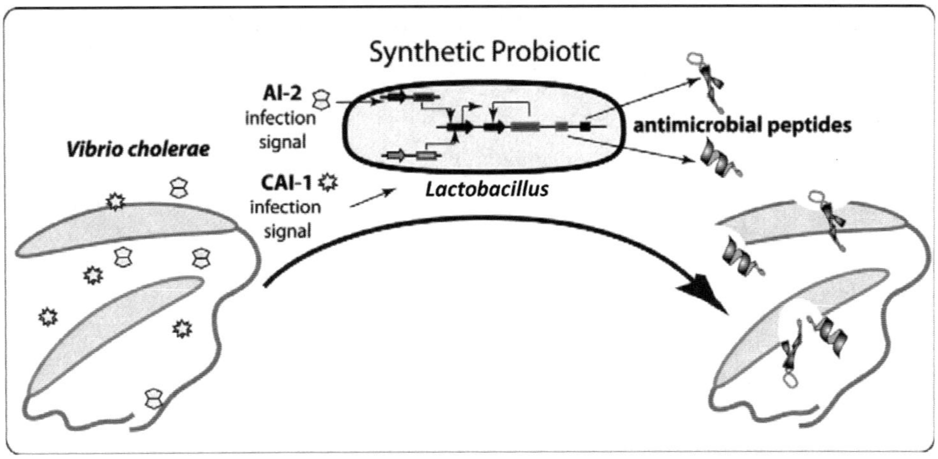

Acknowledgements

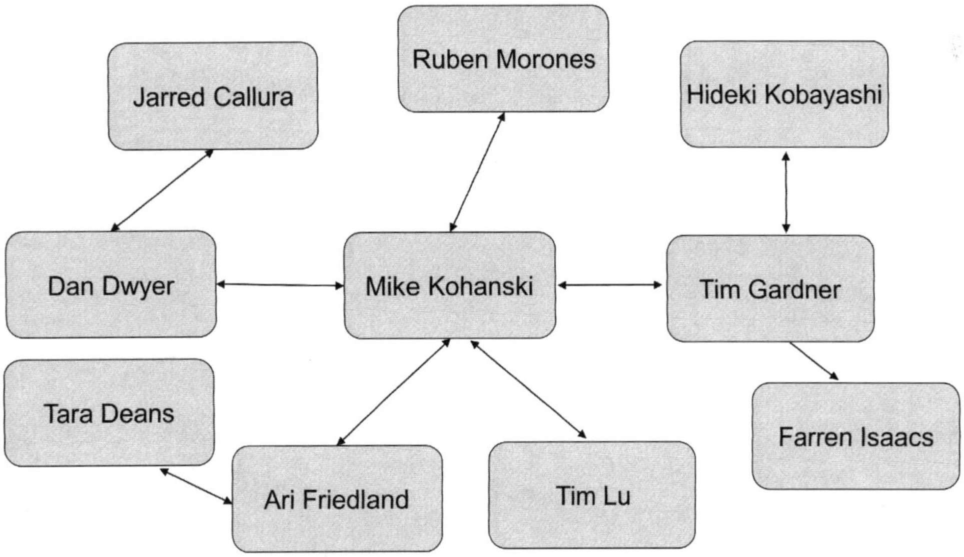

Collins Lab
http://www.bu.edu/abl

Molecular Elucidation and Engineering of Stem Cell Fate Decisions

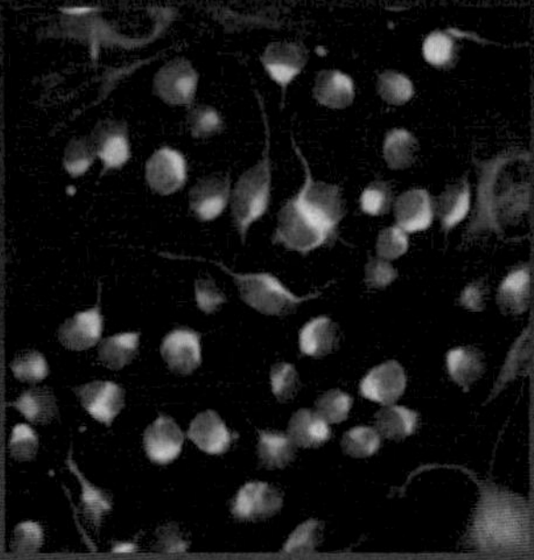

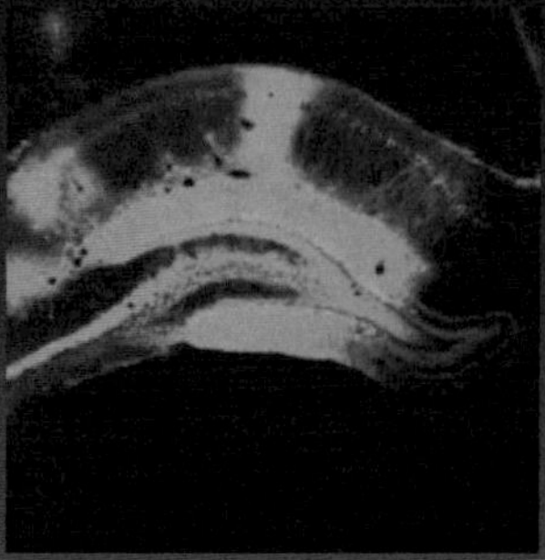

David Schaffer, Ph.D.

Professor, Chemical and Biomolecular Engineering, Bioengineering,
& the Helen Wills Neuroscience Institute
Director, Berkeley Stem Cell Center

University of California at Berkeley

Stem Cell Therapeutics

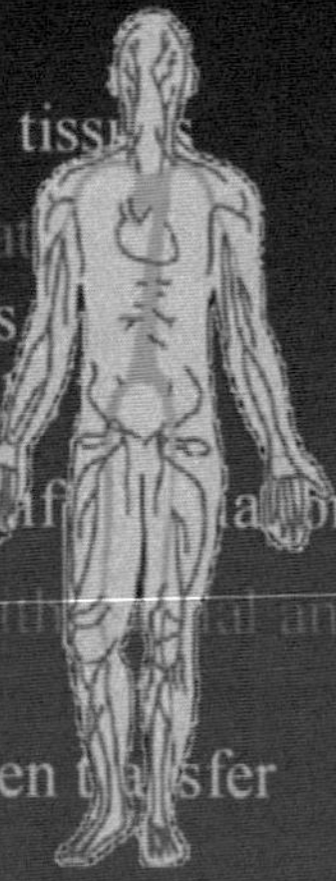

Functional integration into tissue

Low v… engraftment

…unct… …tion of

Diffic… …ring into functional tiss…

Complex and lengthy cell differentia…

- Yi…d heterogeneous cultures
- Co… aminating cells pose ris…

…ll stabil… during expansion and … …a…n

Poorly controlled culture systems with … …al and human-… … proteins

- I… …ogenicity and pathogen t… …sfer
- … …ducibility

Low effi… and r… …ility of stem cell isolation and derivation

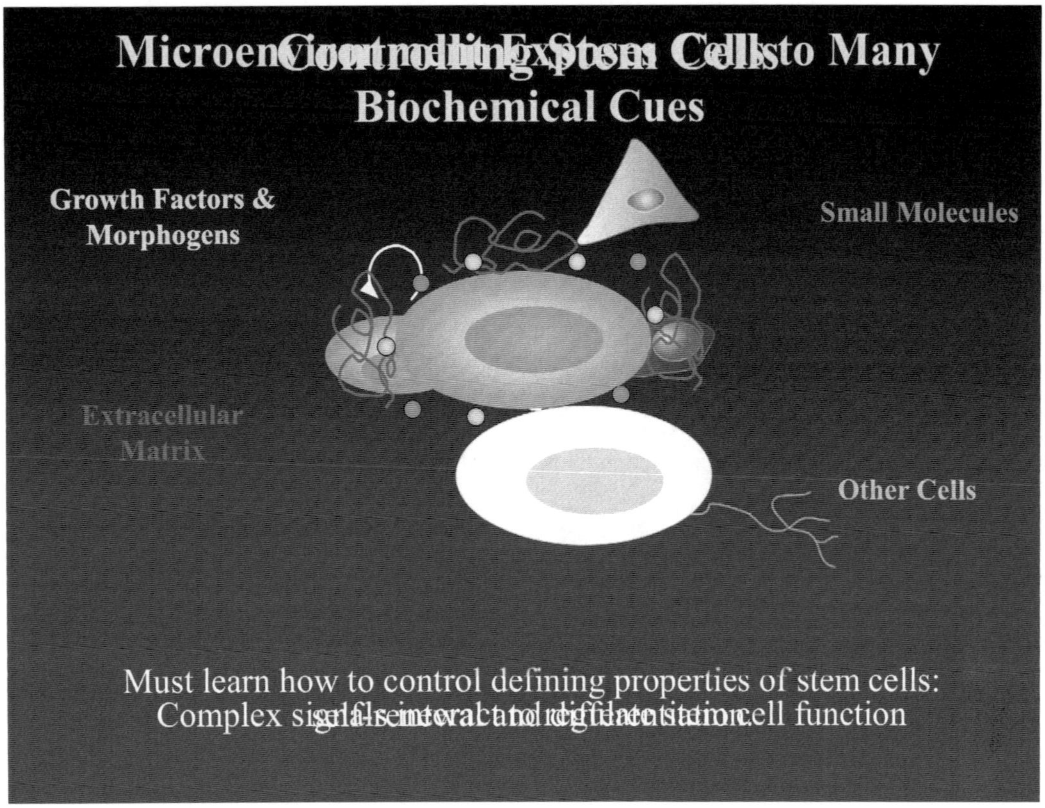
Microenvironment Exposes Cells to Many Biochemical Cues
Controlling Stem Cells
Growth Factors & Morphogens
Small Molecules
Extracellular Matrix
Other Cells
Must learn how to control defining properties of stem cells:
self-renewal and differentiation.
Complex signals interact to regulate stem cell function

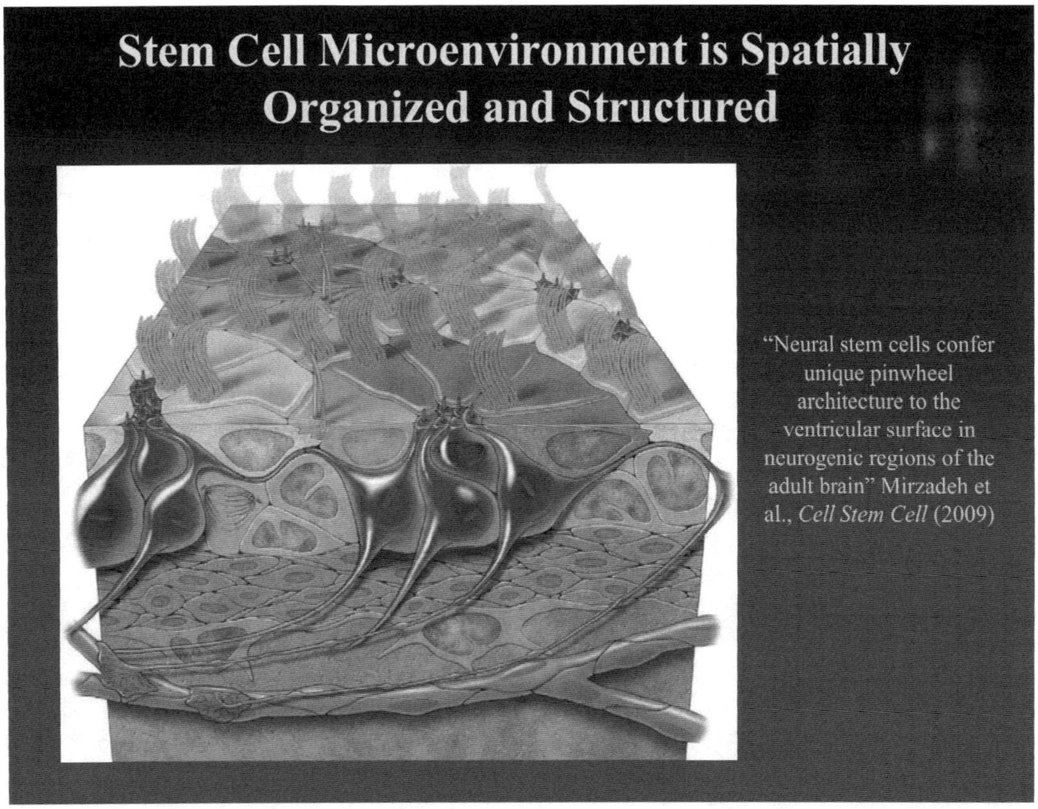
Stem Cell Microenvironment is Spatially Organized and Structured
"Neural stem cells confer unique pinwheel architecture to the ventricular surface in neurogenic regions of the adult brain" Mirzadeh et al., Cell Stem Cell (2009)

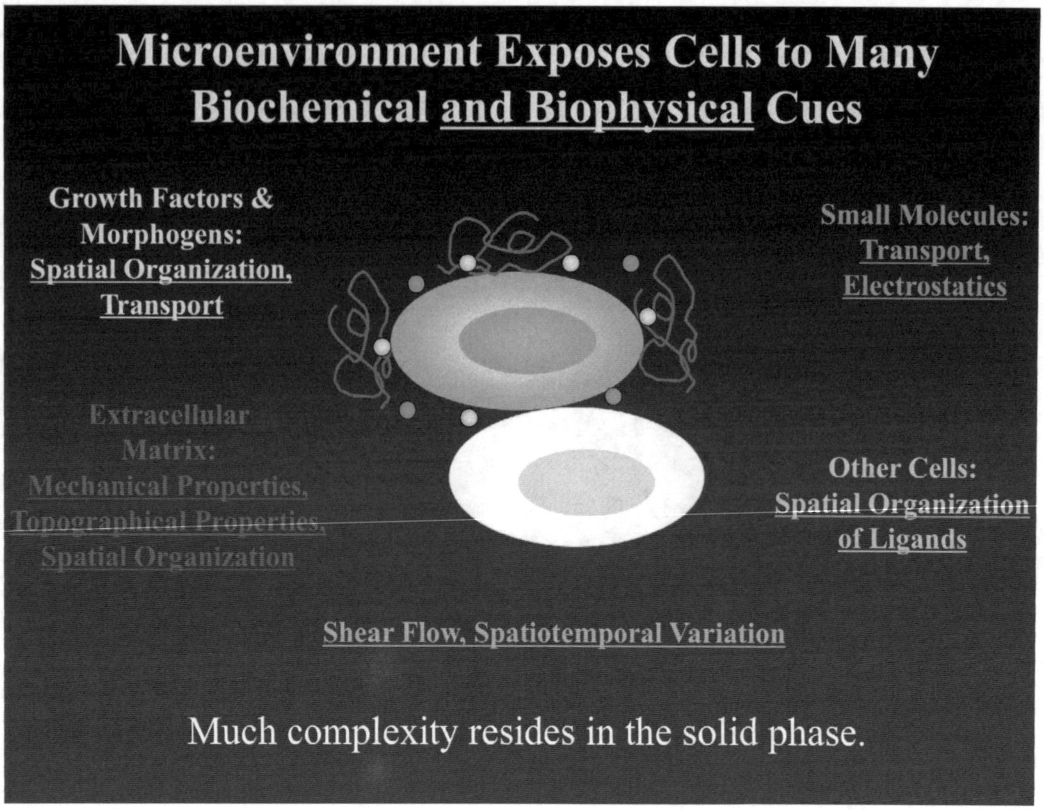
Microenvironment Exposes Cells to Many Biochemical and Biophysical Cues
Growth Factors & Morphogens: Spatial Organization, Transport
Small Molecules: Transport, Electrostatics
Extracellular Matrix: Mechanical Properties, Topographical Properties, Spatial Organization
Other Cells: Spatial Organization of Ligands
Shear Flow, Spatiotemporal Variation
Much complexity resides in the solid phase.

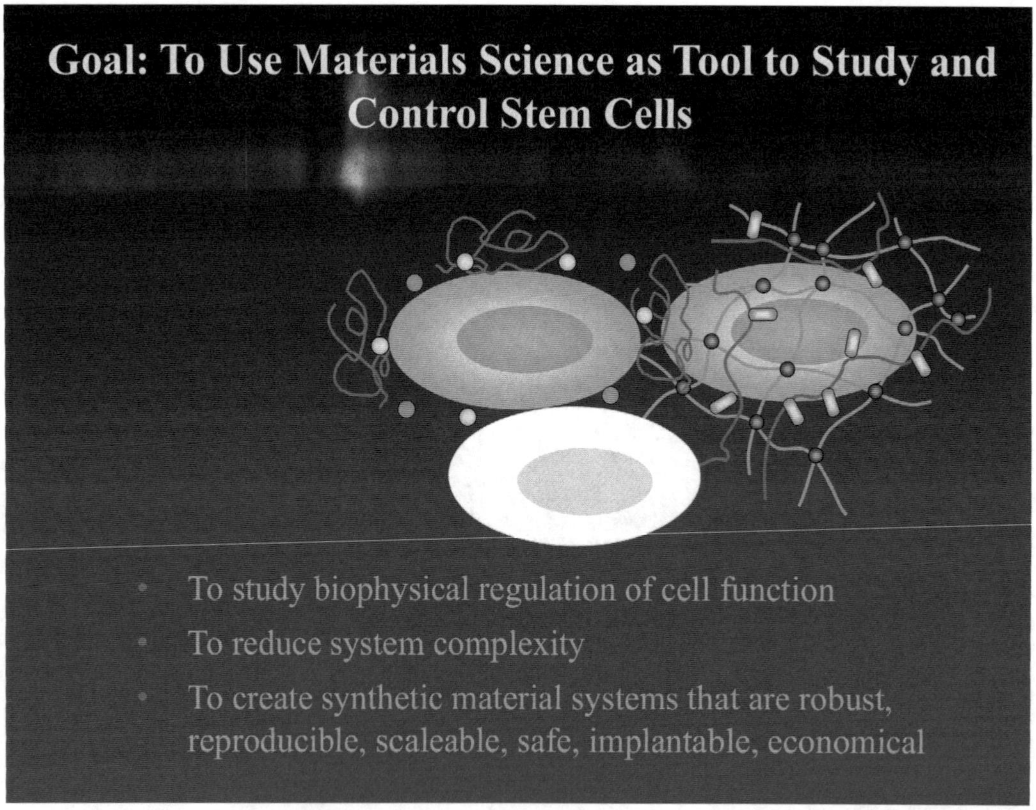
Goal: To Use Materials Science as Tool to Study and Control Stem Cells
To study biophysical regulation of cell function
To reduce system complexity
To create synthetic material systems that are robust, reproducible, scaleable, safe, implantable, economical

Outline

- **Investigating role of mechanics in stem cell function**

- Investigating cell-cell signaling in the stem cell niche

- Biomaterial control over receptor assembly and signaling

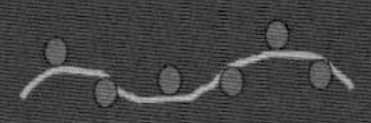

The Adult Mammalian Brain Can Generate New Neurons

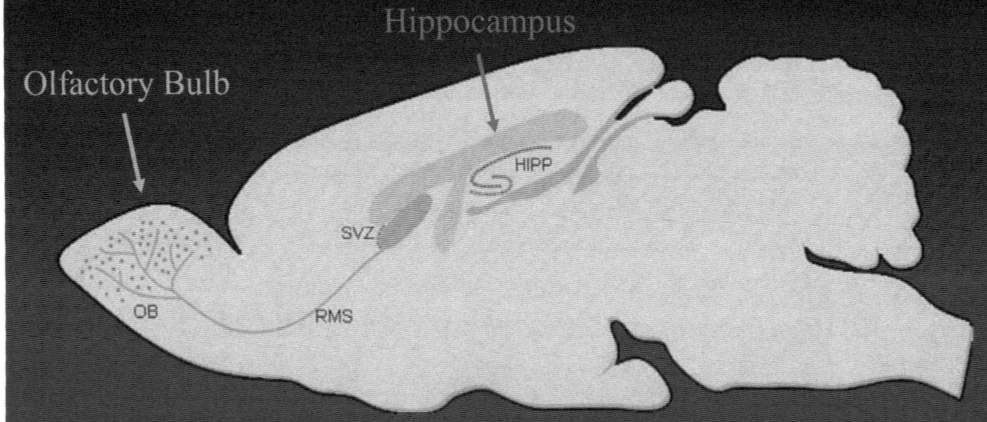

Image from www.scholarpedia.org

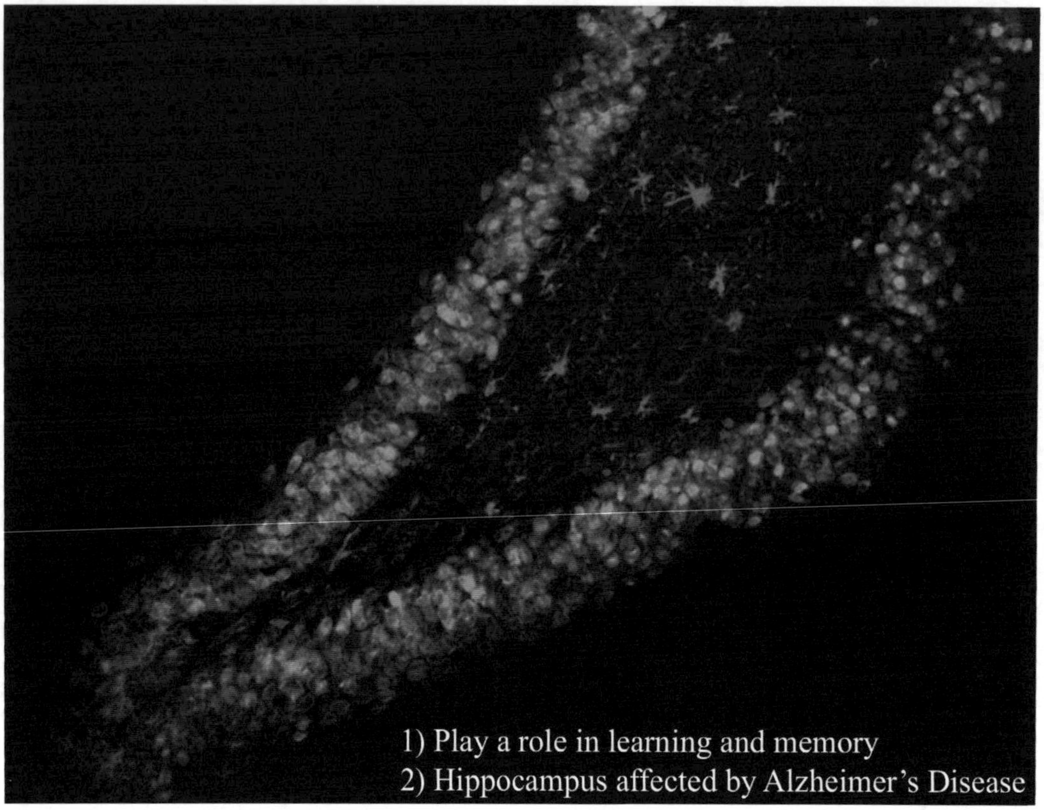
1) Play a role in learning and memory
2) Hippocampus affected by Alzheimer's Disease

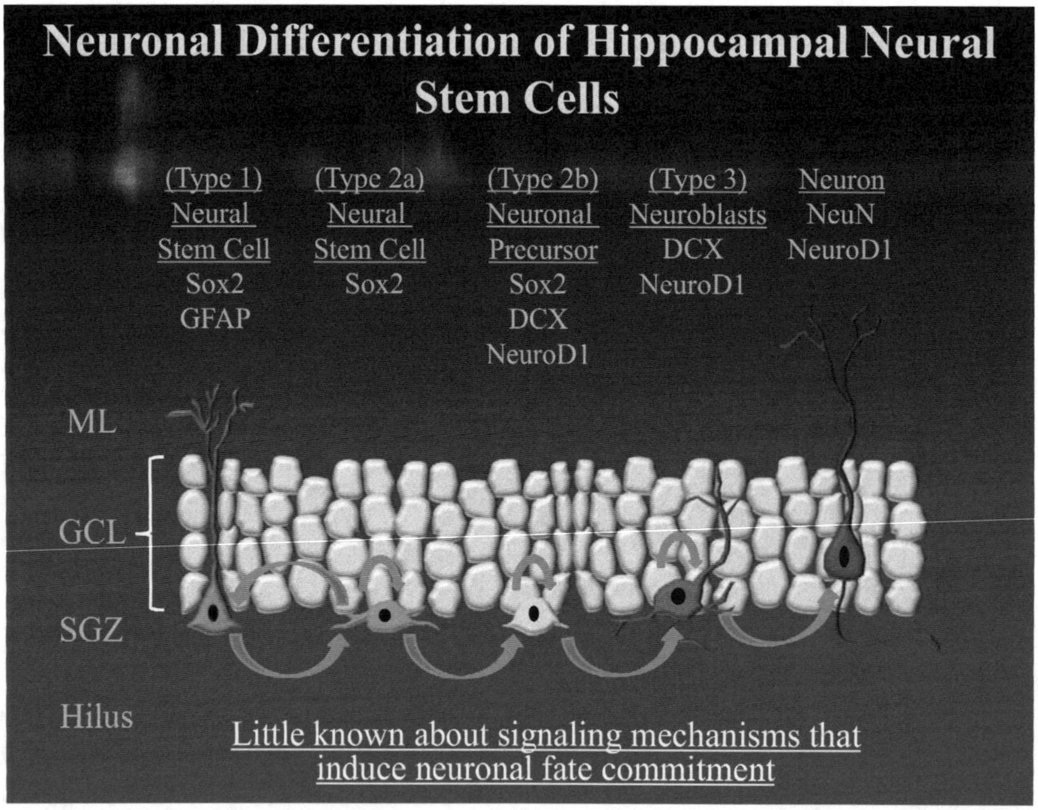
Neuronal Differentiation of Hippocampal Neural Stem Cells
(Type 1) Neural Stem Cell Sox2 GFAP
(Type 2a) Neural Stem Cell Sox2
(Type 2b) Neuronal Precursor Sox2 DCX NeuroD1
(Type 3) Neuroblasts DCX NeuroD1
Neuron NeuN NeuroD1
ML
GCL
SGZ
Hilus
Little known about signaling mechanisms that induce neuronal fate commitment

Adult Neural Stem Cells are Multipotent in Culture

Neurons | Astrocytes | Oligodendrocytes

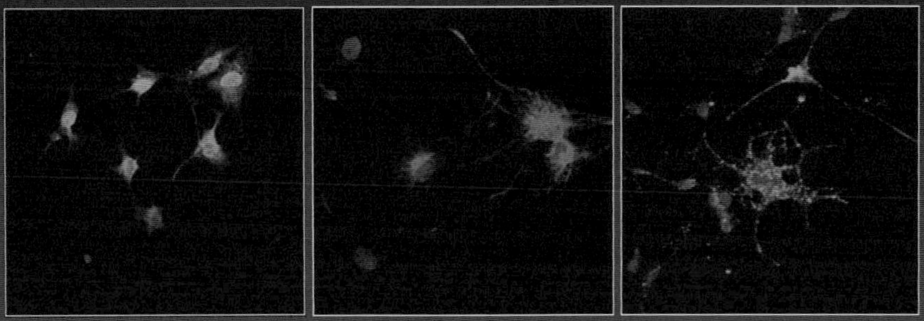

Blue = nuclei

Range of Cell Substrate Elasticity

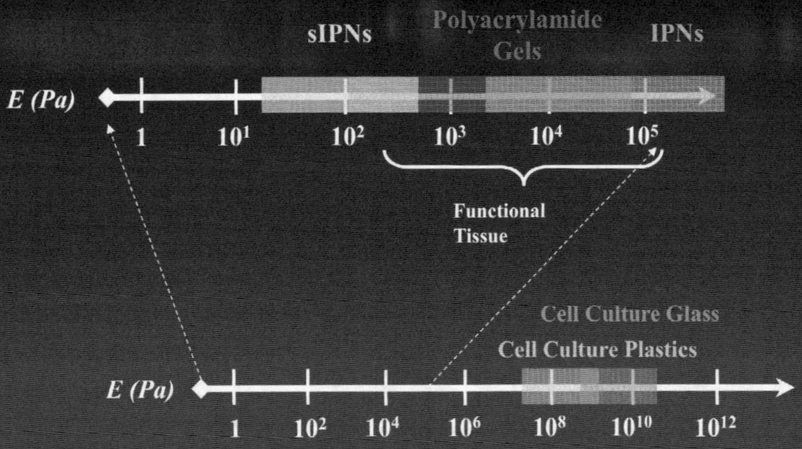

Does *E* affect cell self-renewal and/or differentiation?

Sheu MT, *et al.* 2001. *Biomaterials* 22(13):1713 & Engler/Discher, UPenn

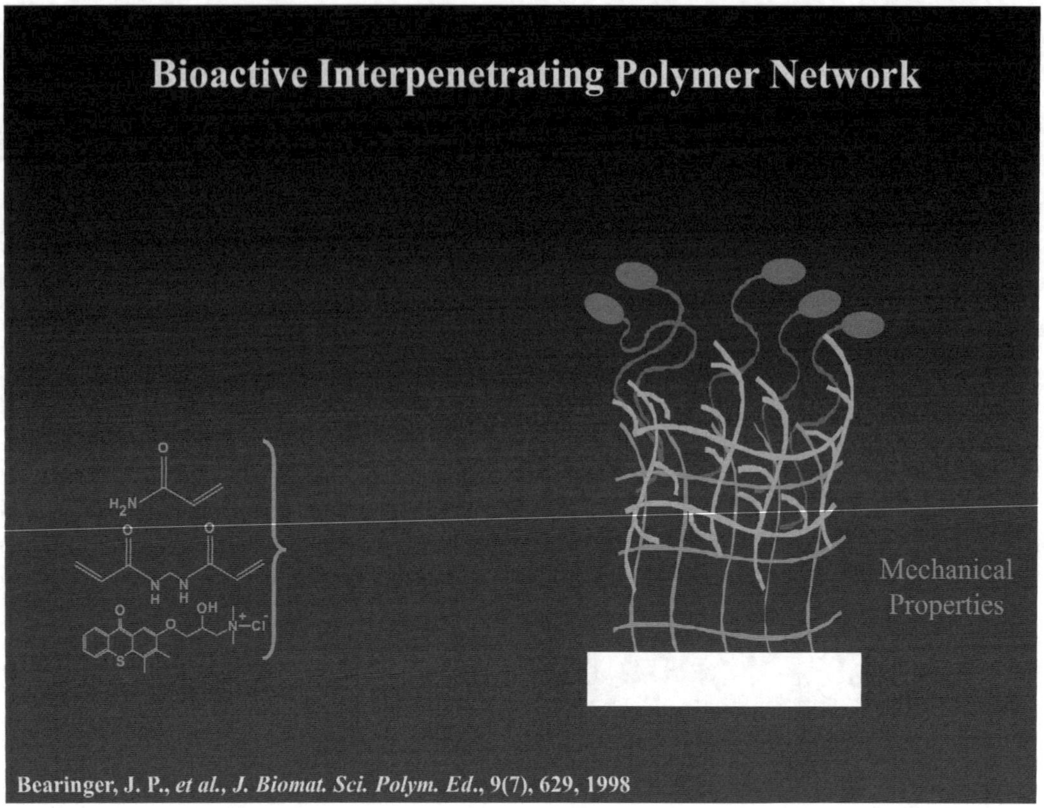
Bioactive Interpenetrating Polymer Network
Mechanical
Properties
Bearinger, J. P., et al., J. Biomat. Sci. Polym. Ed., 9(7), 629, 1998

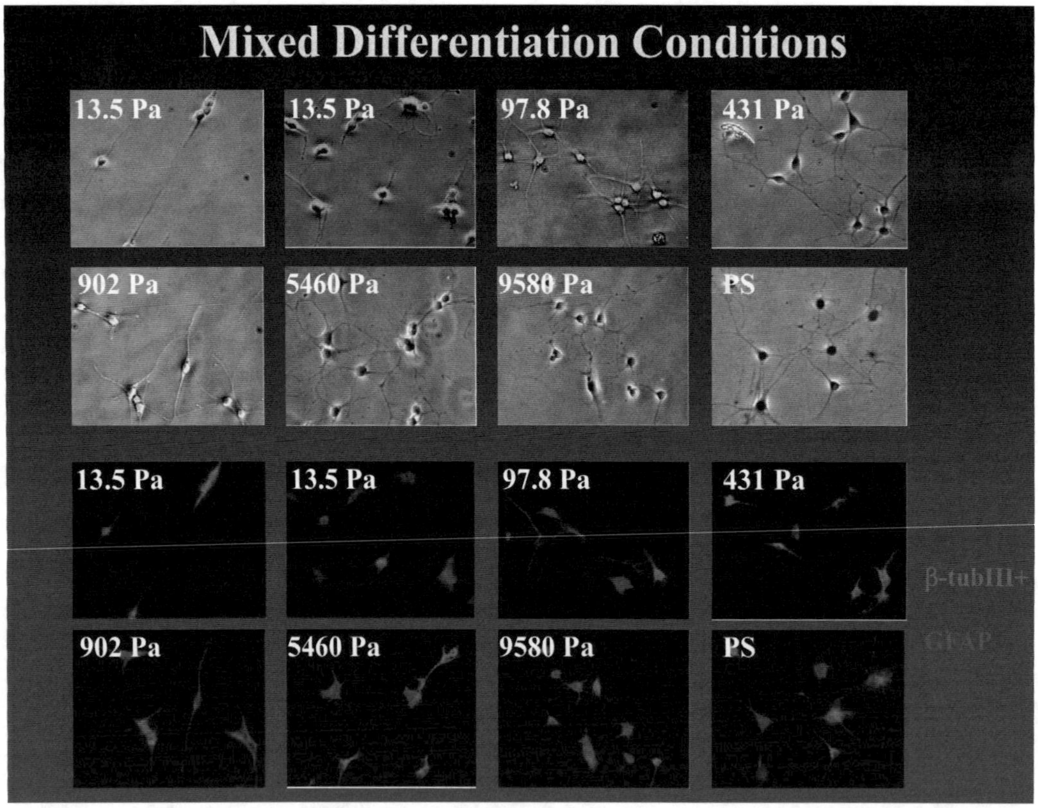
Mixed Differentiation Conditions
13.5 Pa
13.5 Pa
97.8 Pa
431 Pa
902 Pa
5460 Pa
9580 Pa
PS
13.5 Pa
13.5 Pa
97.8 Pa
431 Pa
902 Pa
5460 Pa
9580 Pa
PS
β-tubIII+
GFAP

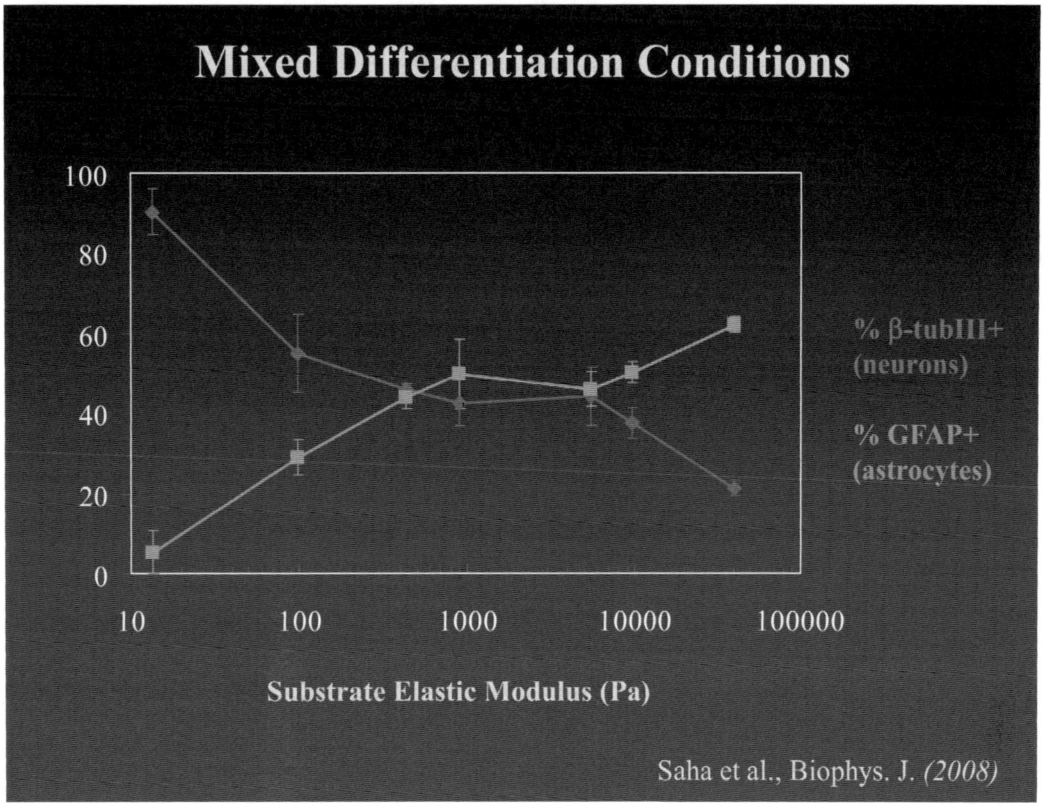
Mixed Differentiation Conditions
100
80
60
40
20
0
10
100
1000
10000
100000
Substrate Elastic Modulus (Pa)
% β-tubIII+ (neurons)
% GFAP+ (astrocytes)
Saha et al., Biophys. J. (2008)

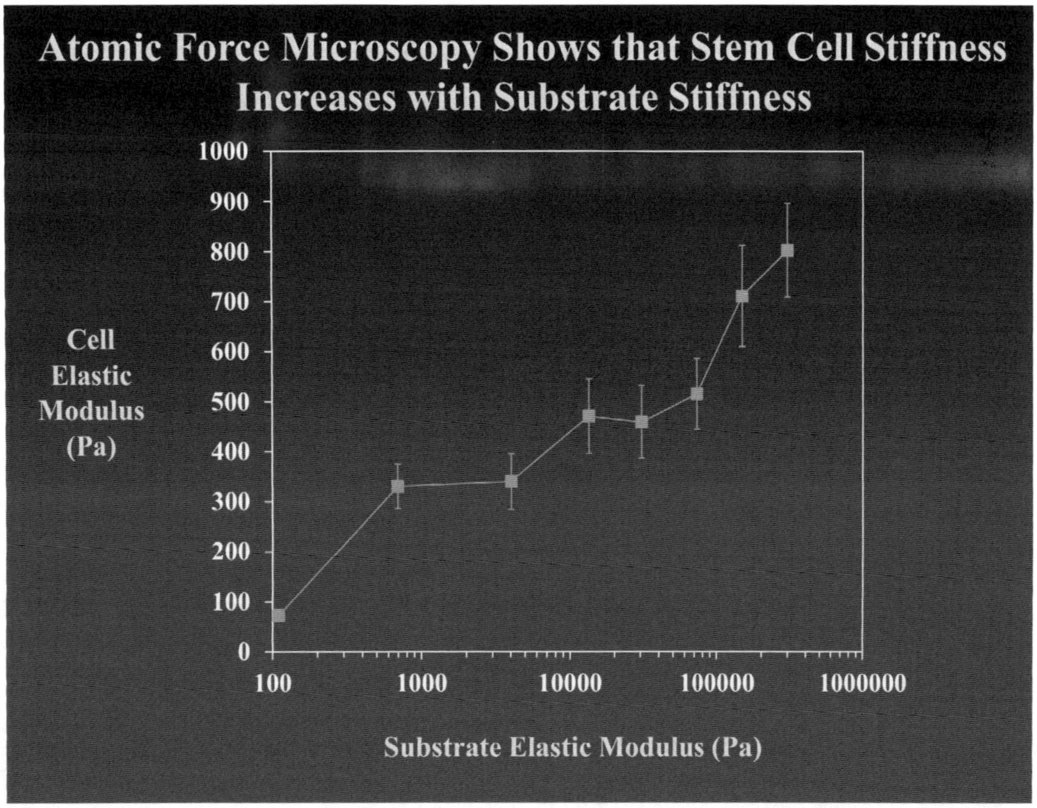
Atomic Force Microscopy Shows that Stem Cell Stiffness Increases with Substrate Stiffness
Cell Elastic Modulus (Pa)
1000
900
800
700
600
500
400
300
200
100
0
100
1000
10000
100000
1000000
Substrate Elastic Modulus (Pa)

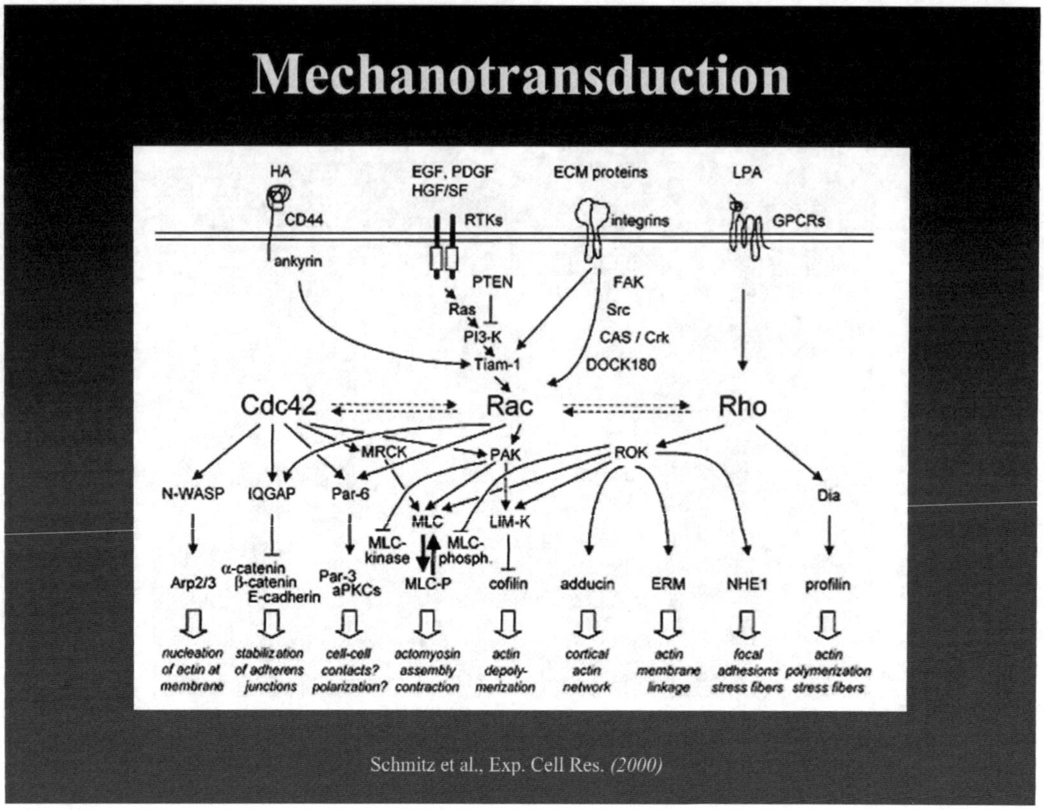
Mechanotransduction
HA
CD44
ankyrin
EGF, PDGF
HGF/SF
RTKs
PTEN
Ras
PI3-K
Tiam-1
ECM proteins
integrins
FAK
Src
CAS / Crk
DOCK180
LPA
GPCRs
Cdc42
Rac
Rho
MRCK
PAK
ROK
N-WASP
IQGAP
Par-6
MLC
LIM-K
Dia
MLC-kinase
MLC-phosph.
Arp2/3
α-catenin
β-catenin
E-cadherin
Par-3
aPKCs
MLC-P
cofilin
adducin
ERM
NHE1
profilin
nucleation of actin at membrane
stabilization of adherens junctions
cell-cell contacts? polarization?
actomyosin assembly contraction
actin depoly-merization
cortical actin network
actin membrane linkage
focal adhesions stress fibers
actin polymerization stress fibers
Schmitz et al., Exp. Cell Res. (2000)

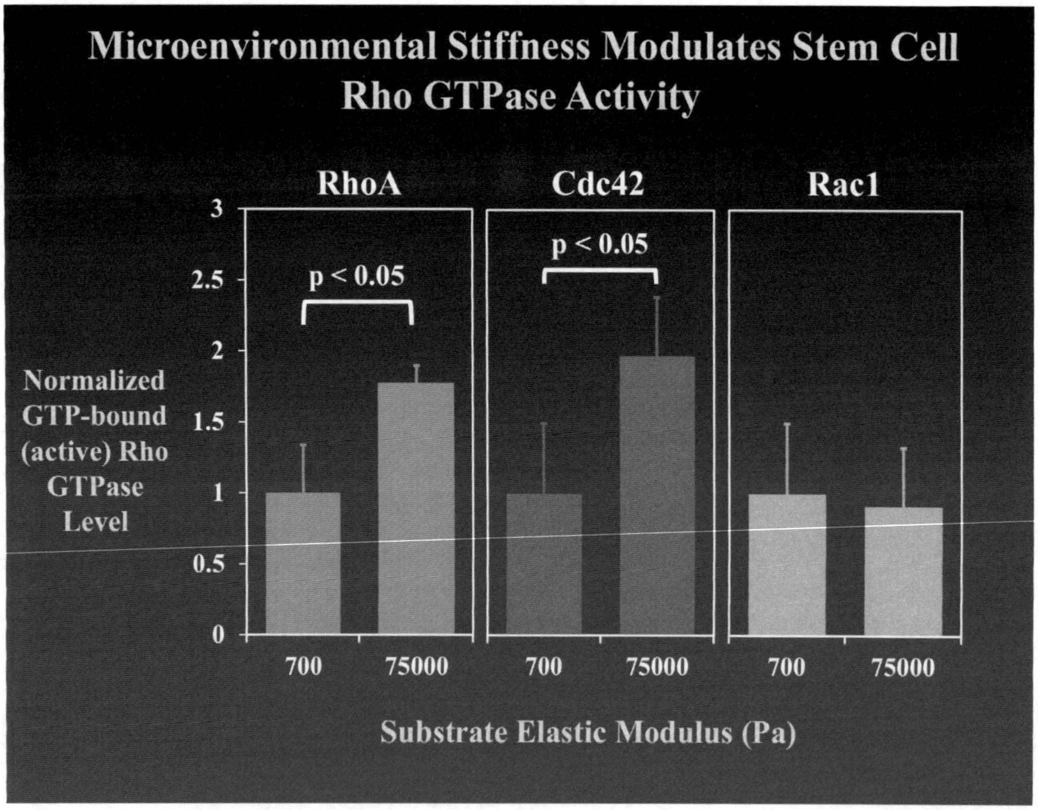
Microenvironmental Stiffness Modulates Stem Cell Rho GTPase Activity
RhoA
Cdc42
Rac1
p < 0.05
p < 0.05
Normalized GTP-bound (active) Rho GTPase Level
0
0.5
1
1.5
2
2.5
3
700
75000
700
75000
700
75000
Substrate Elastic Modulus (Pa)

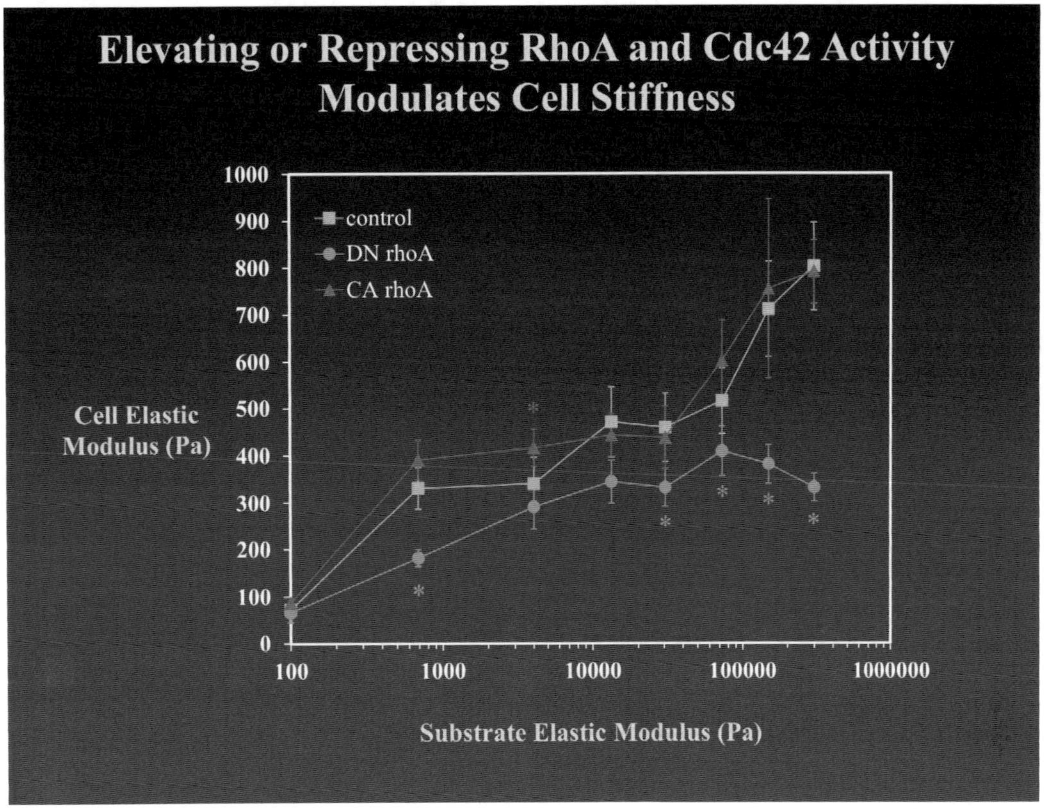
Elevating or Repressing RhoA and Cdc42 Activity Modulates Cell Stiffness
control
DN rhoA
CA rhoA
Cell Elastic Modulus (Pa)
Substrate Elastic Modulus (Pa)
0
100
200
300
400
500
600
700
800
900
1000
100
1000
10000
100000
1000000

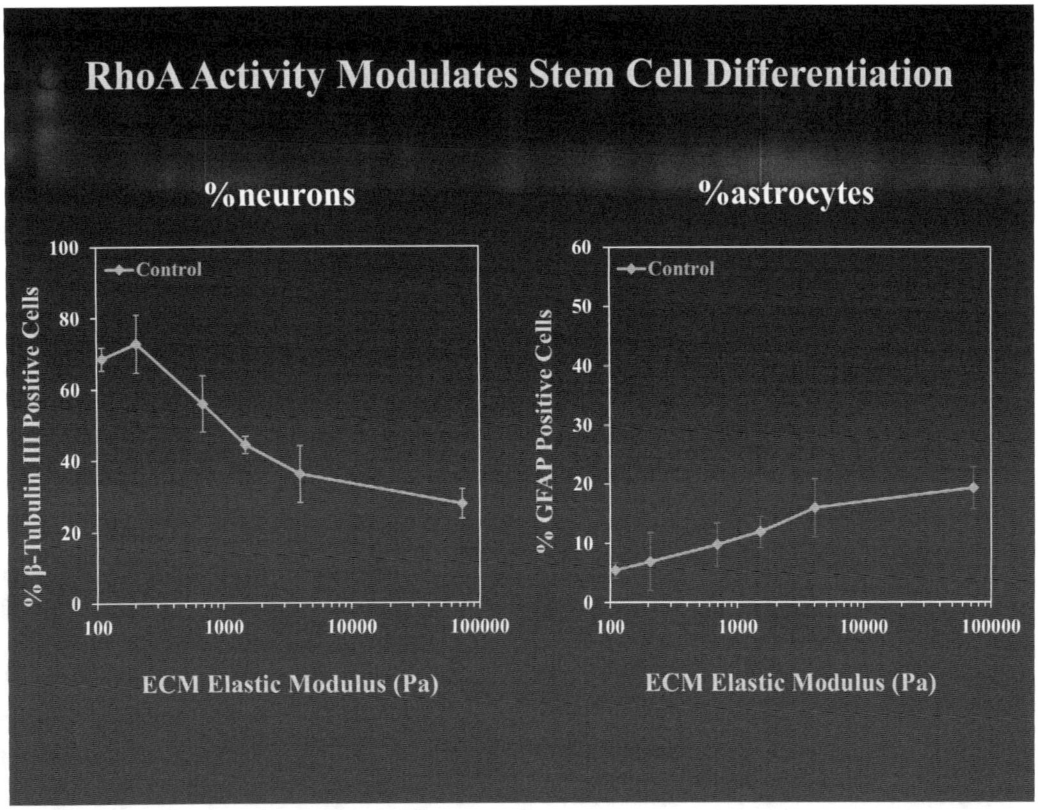
RhoA Activity Modulates Stem Cell Differentiation
%neurons
%astrocytes
Control
% β-Tubulin III Positive Cells
% GFAP Positive Cells
ECM Elastic Modulus (Pa)
ECM Elastic Modulus (Pa)
0
20
40
60
80
100
0
10
20
30
40
50
60
100
1000
10000
100000

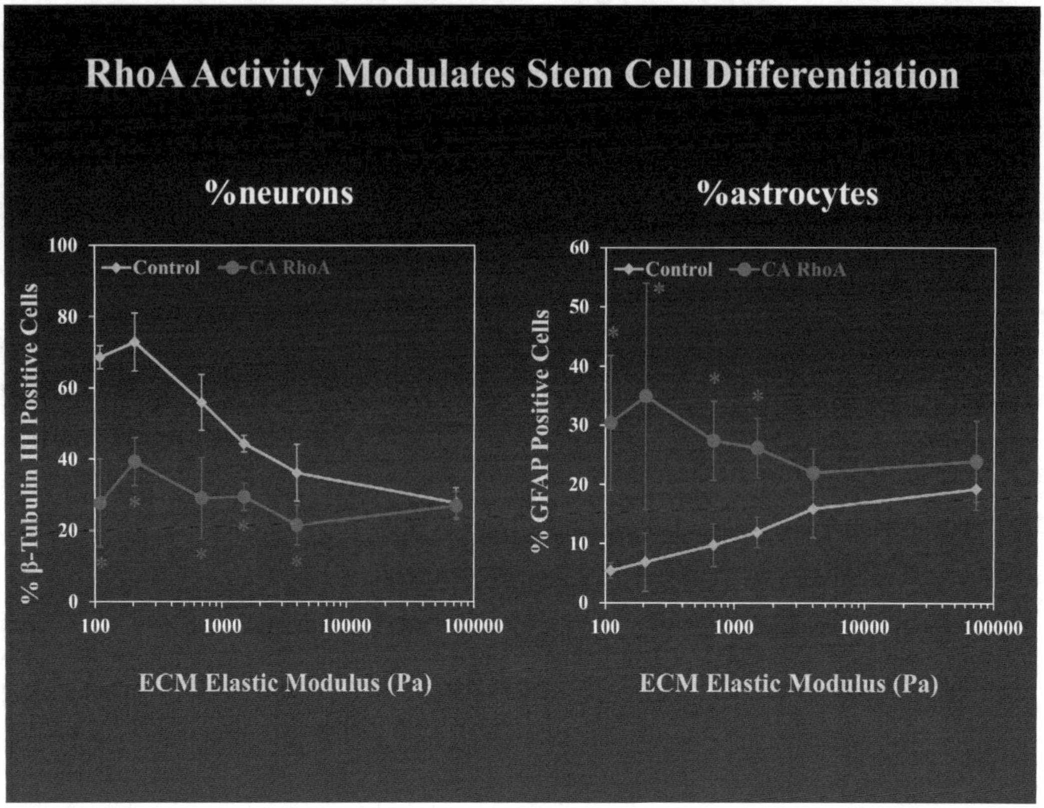
RhoA Activity Modulates Stem Cell Differentiation
%neurons
%astrocytes
Control
CA RhoA
% β-Tubulin III Positive Cells
% GFAP Positive Cells
ECM Elastic Modulus (Pa)
ECM Elastic Modulus (Pa)

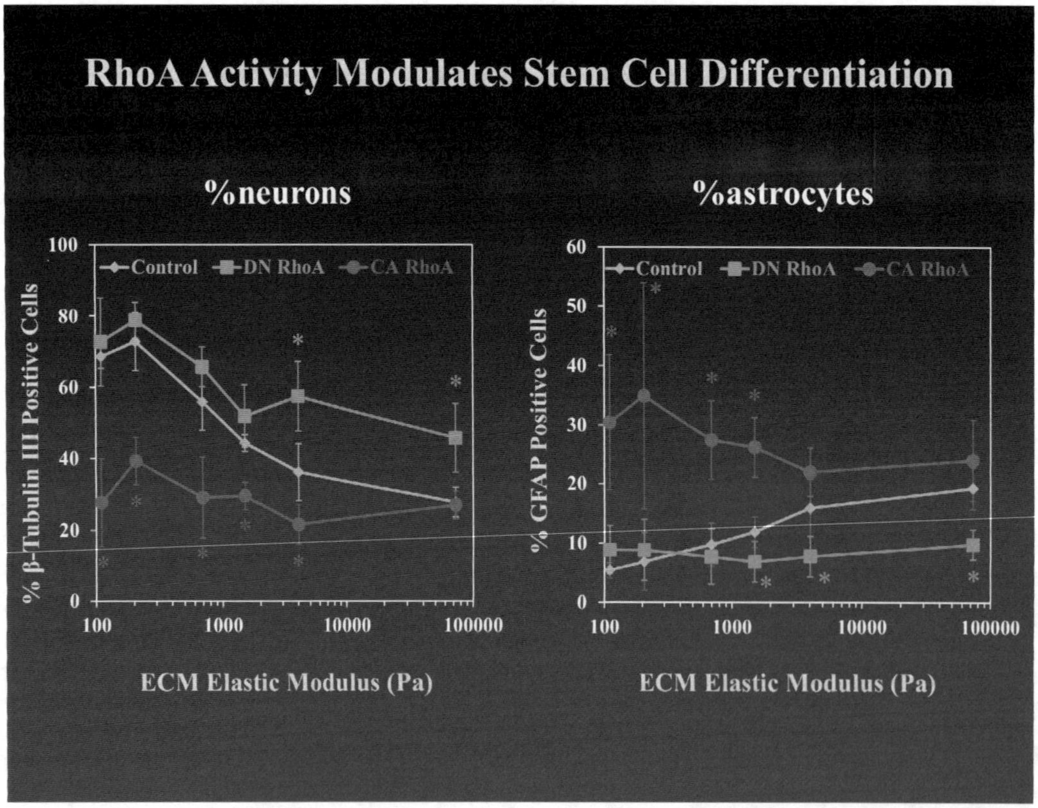
RhoA Activity Modulates Stem Cell Differentiation
%neurons
%astrocytes
Control
DN RhoA
CA RhoA
% β-Tubulin III Positive Cells
% GFAP Positive Cells
ECM Elastic Modulus (Pa)
ECM Elastic Modulus (Pa)

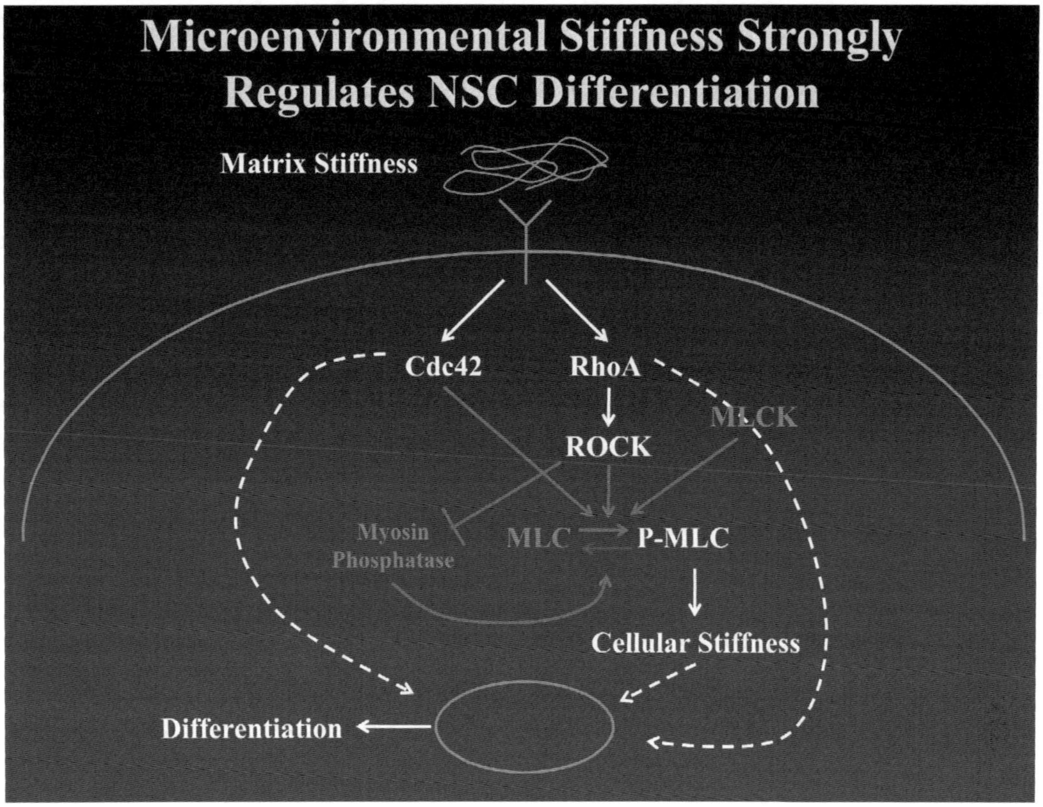
Microenvironmental Stiffness Strongly Regulates NSC Differentiation
Matrix Stiffness
Cdc42
RhoA
ROCK
MLCK
Myosin Phosphatase
MLC
P-MLC
Cellular Stiffness
Differentiation

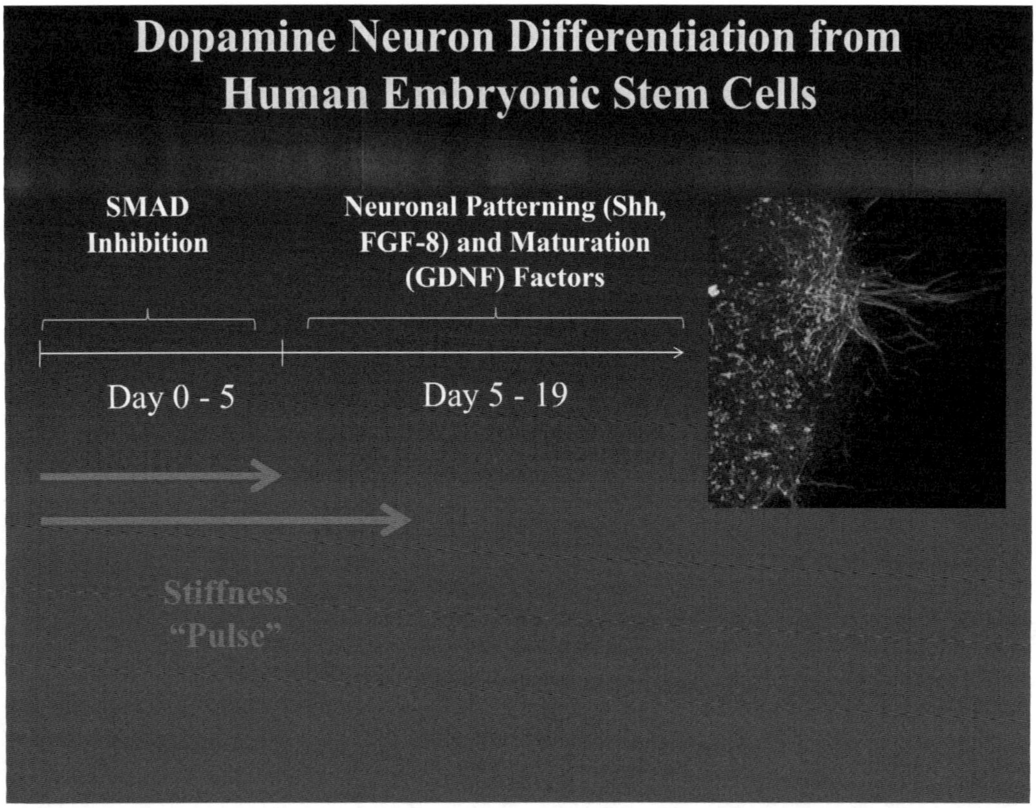
Dopamine Neuron Differentiation from Human Embryonic Stem Cells
SMAD Inhibition
Neuronal Patterning (Shh, FGF-8) and Maturation (GDNF) Factors
Day 0 - 5
Day 5 - 19
Stiffness "Pulse"

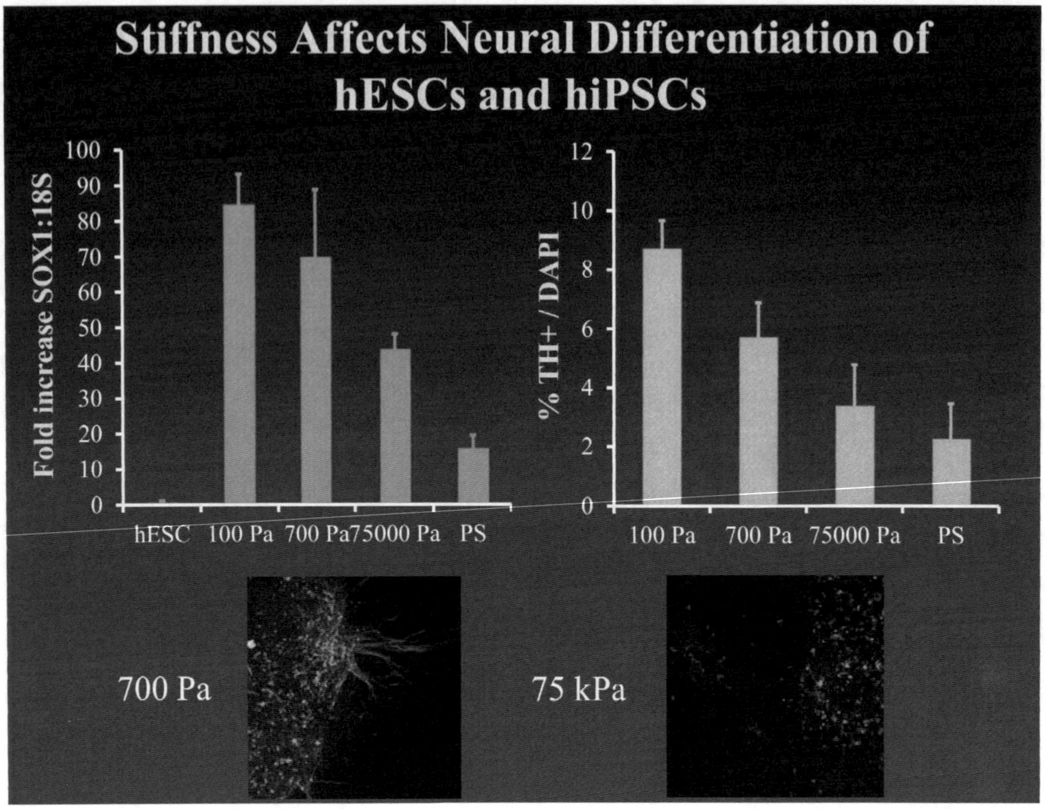
Stiffness Affects Neural Differentiation of hESCs and hiPSCs
Fold increase SOX1:18S
100
90
80
70
60
50
40
30
20
10
0
hESC
100 Pa
700 Pa
75000 Pa
PS
% TH+ / DAPI
12
10
8
6
4
2
0
100 Pa
700 Pa
75000 Pa
PS
700 Pa
75 kPa

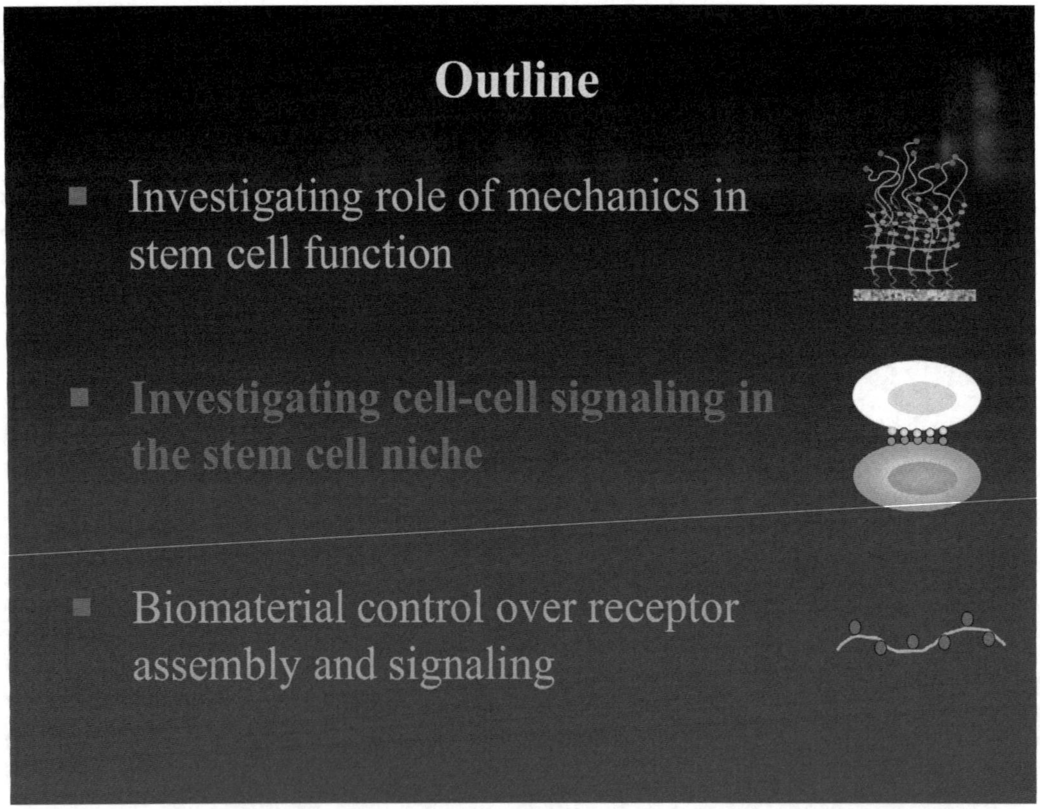
Outline
Investigating role of mechanics in stem cell function
Investigating cell-cell signaling in the stem cell niche
Biomaterial control over receptor assembly and signaling

Eph/Ephrins

- Ephs are cell surface receptors, and Ephrins are cell-associated ligands
- Active only when clustered
- Signaling system implicated in numerous developmental processes, and particularly axon pathfinding in the CNS
- Signaling is typically bidirectional
- Signal via numerous pathways, most notably ones involved in cytoskeletal rearrangements

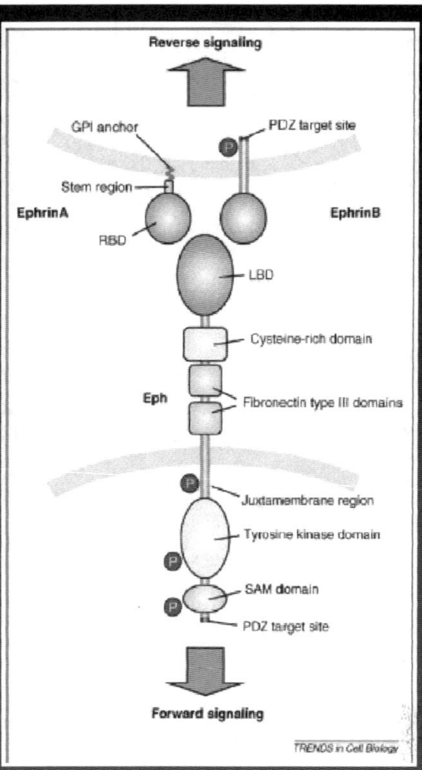

Egea *et al* (2005)

Ephrin-B2 and EphB4 Expression in the Brain

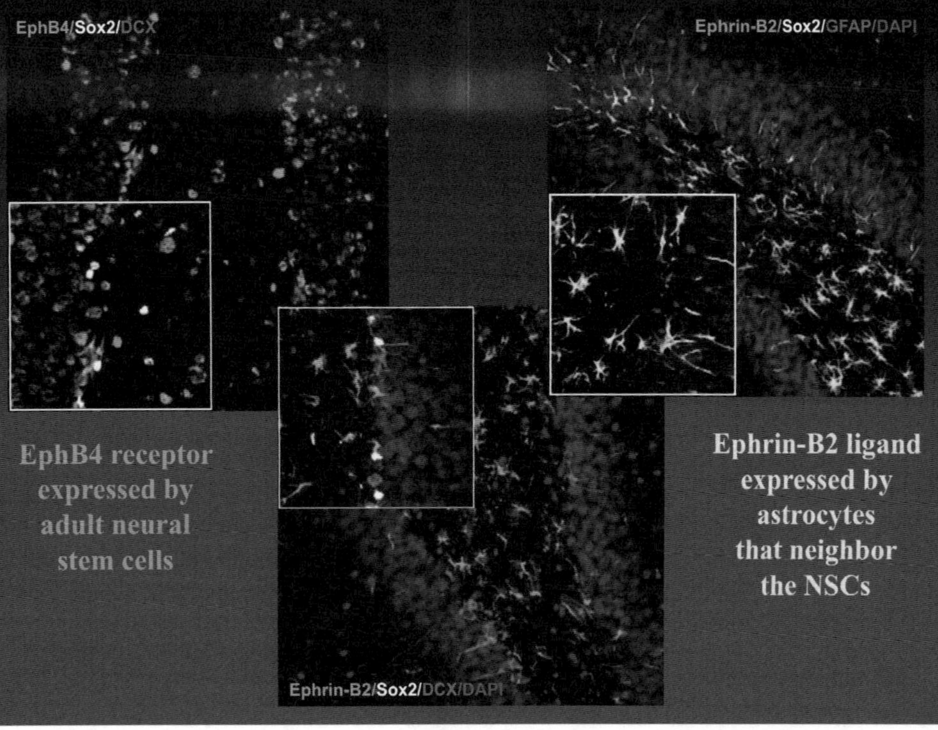

EphB4 receptor expressed by adult neural stem cells

Ephrin-B2 ligand expressed by astrocytes that neighbor the NSCs

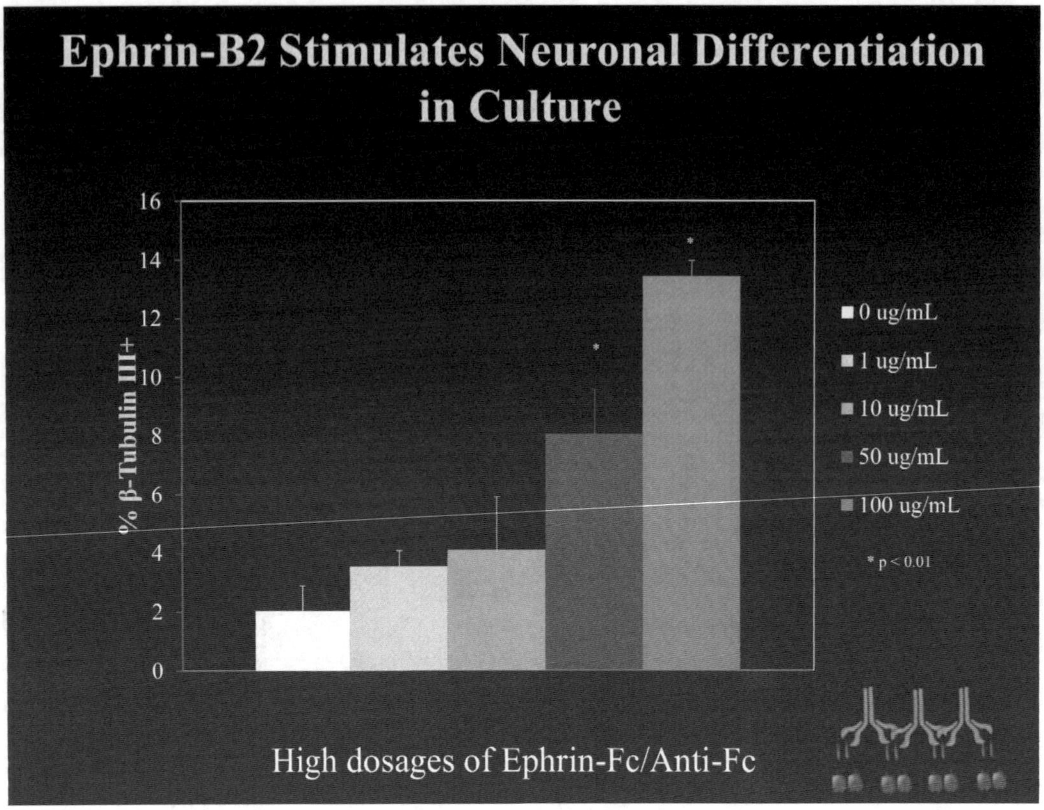
Ephrin-B2 Stimulates Neuronal Differentiation in Culture
% β-Tubulin III+
16
14
12
10
8
6
4
2
0
*
*
0 ug/mL
1 ug/mL
10 ug/mL
50 ug/mL
100 ug/mL
* p < 0.01
High dosages of Ephrin-Fc/Anti-Fc

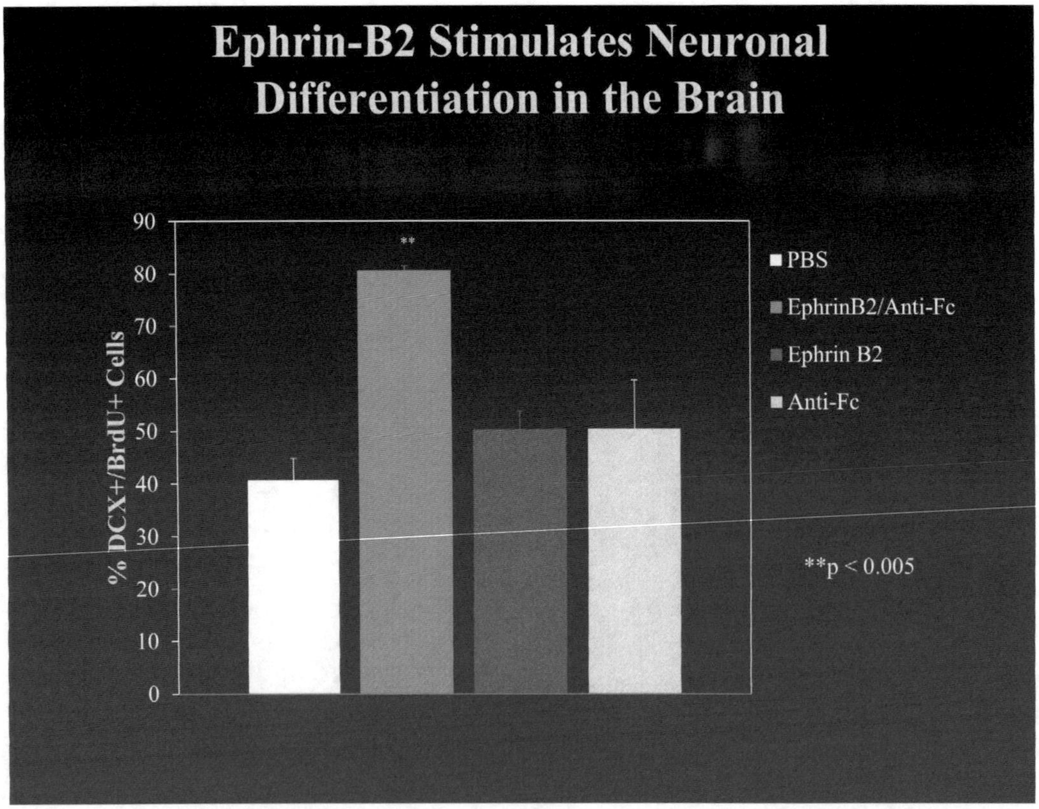
Ephrin-B2 Stimulates Neuronal Differentiation in the Brain
% DCX+/BrdU+ Cells
90
80
70
60
50
40
30
20
10
0
**
PBS
EphrinB2/Anti-Fc
Ephrin B2
Anti-Fc
**p < 0.005

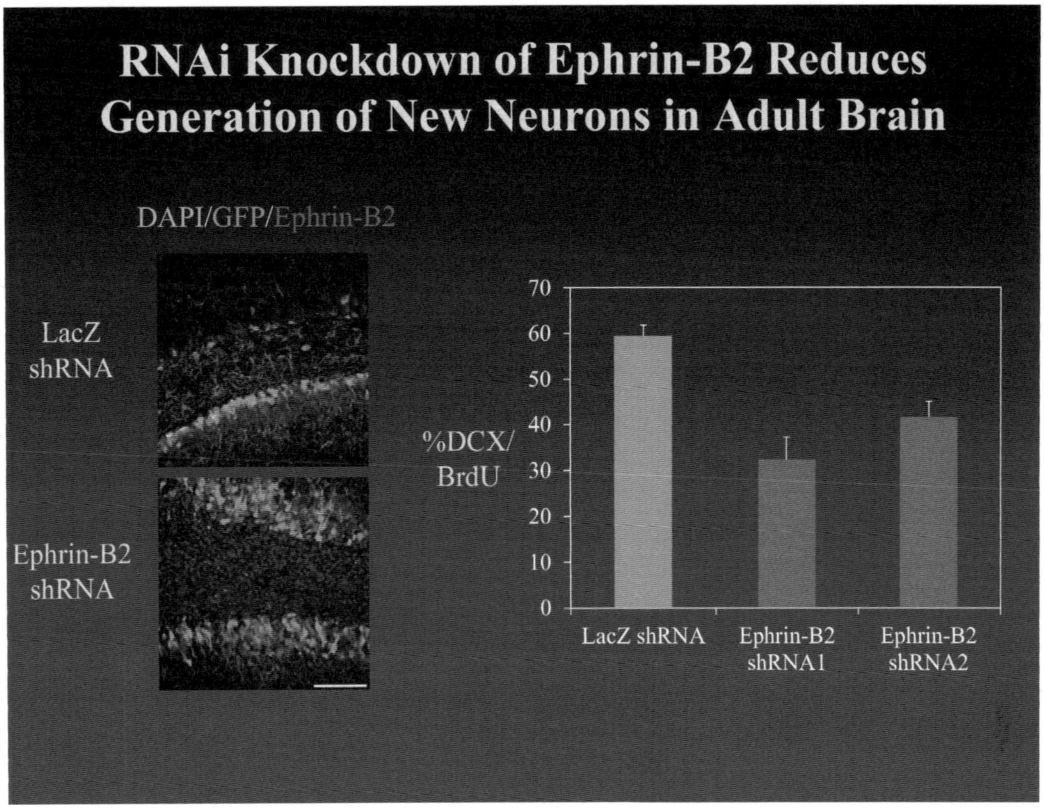
RNAi Knockdown of Ephrin-B2 Reduces Generation of New Neurons in Adult Brain
DAPI/GFP/Ephrin-B2
LacZ shRNA
Ephrin-B2 shRNA
%DCX/ BrdU
70
60
50
40
30
20
10
0
LacZ shRNA
Ephrin-B2 shRNA1
Ephrin-B2 shRNA2

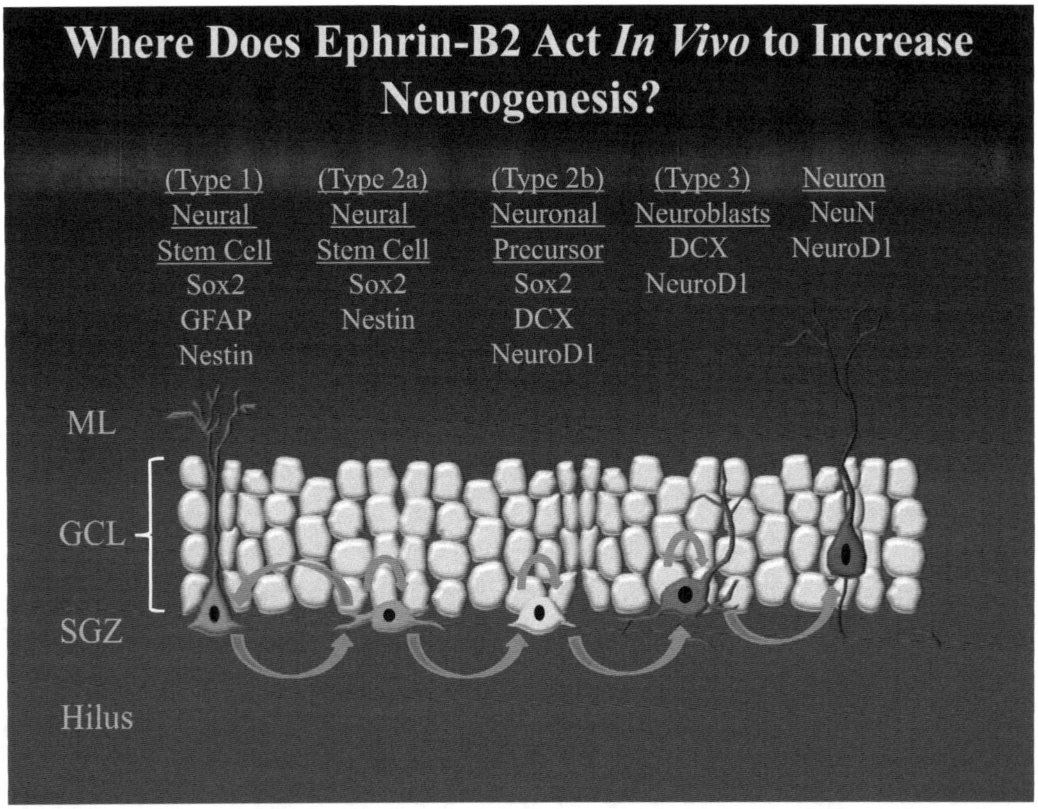
Where Does Ephrin-B2 Act In Vivo to Increase Neurogenesis?
(Type 1) Neural Stem Cell
Sox2
GFAP
Nestin
(Type 2a) Neural Stem Cell
Sox2
Nestin
(Type 2b) Neuronal Precursor
Sox2
DCX
NeuroD1
(Type 3) Neuroblasts
DCX
NeuroD1
Neuron
NeuN
NeuroD1
ML
GCL
SGZ
Hilus

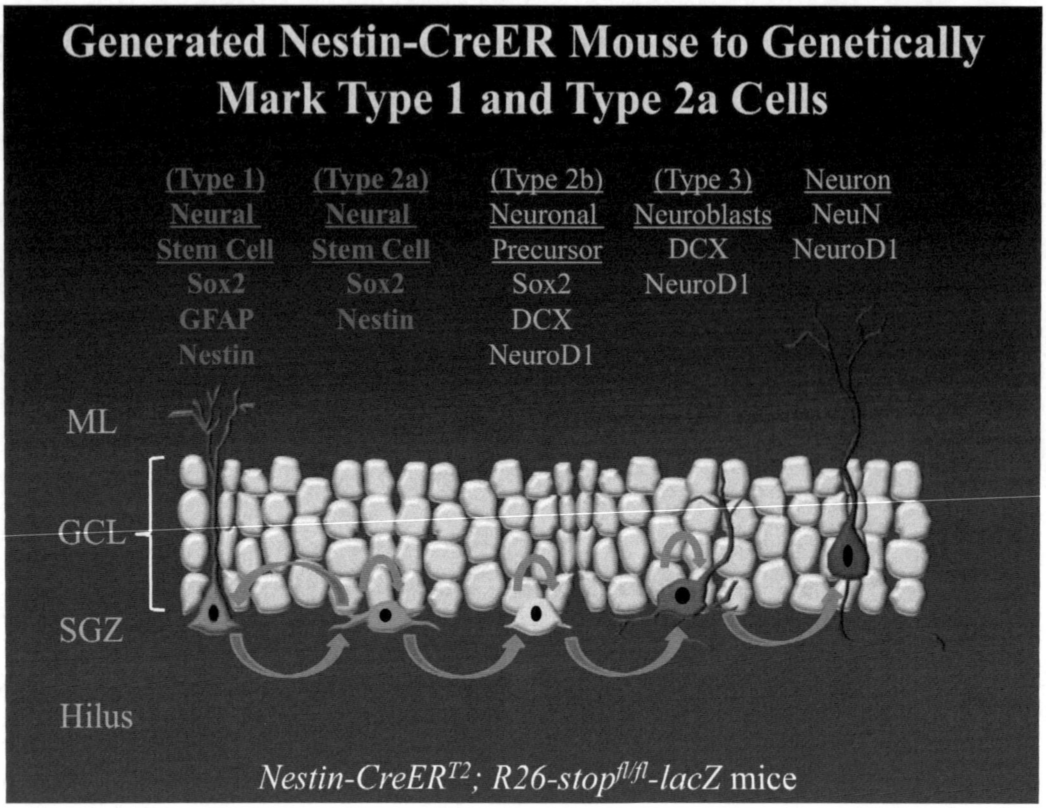
Generated Nestin-CreER Mouse to Genetically Mark Type 1 and Type 2a Cells
(Type 1)
Neural
Stem Cell
Sox2
GFAP
Nestin
(Type 2a)
Neural
Stem Cell
Sox2
Nestin
(Type 2b)
Neuronal
Precursor
Sox2
DCX
NeuroD1
(Type 3)
Neuroblasts
DCX
NeuroD1
Neuron
NeuN
NeuroD1
ML
GCL
SGZ
Hilus
Nestin-CreERT2; R26-stopfl/fl-lacZ mice

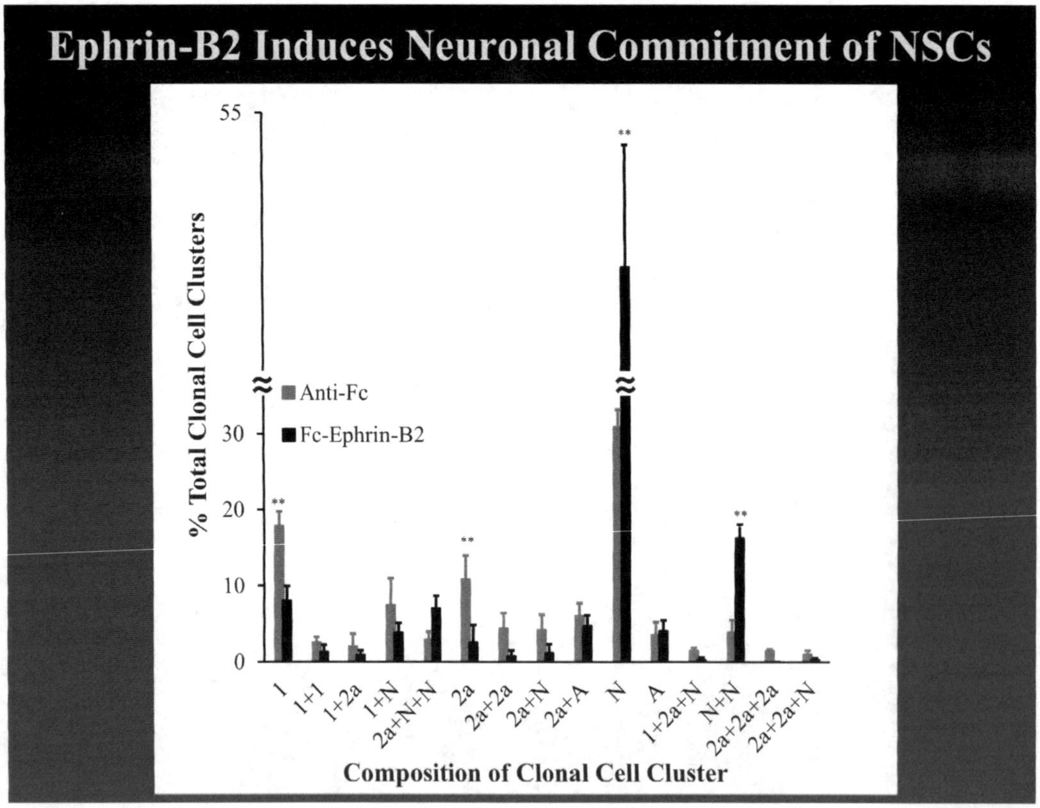
Ephrin-B2 Induces Neuronal Commitment of NSCs
% Total Clonal Cell Clusters
55
30
20
10
0
Anti-Fc
Fc-Ephrin-B2
1
1+1
1+2a
1+N
2a+N+N
2a
2a+2a
2a+N
2a+A
N
A
1+2a+N
N+N
2a+2a+2a
2a+2a+N
Composition of Clonal Cell Cluster

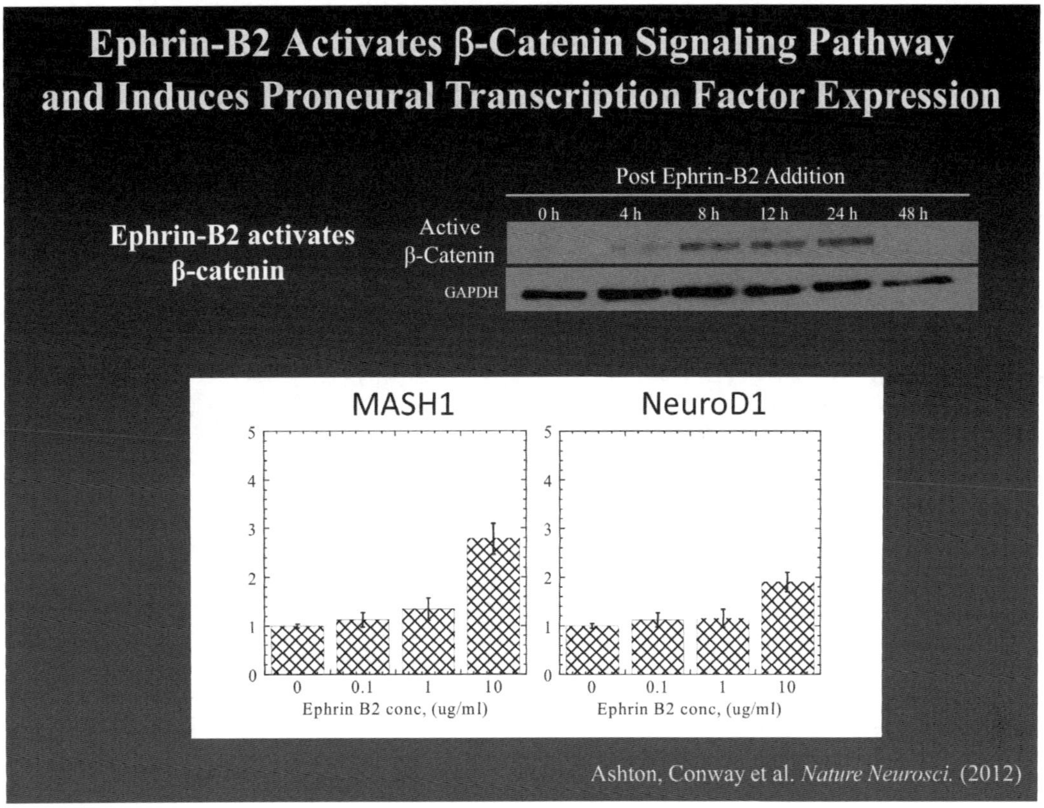
Ephrin-B2 Activates β-Catenin Signaling Pathway and Induces Proneural Transcription Factor Expression
Ephrin-B2 activates β-catenin
Post Ephrin-B2 Addition
0 h
4 h
8 h
12 h
24 h
48 h
Active β-Catenin
GAPDH
MASH1
NeuroD1
5
4
3
2
1
0
0
0.1
1
10
Ephrin B2 conc, (ug/ml)
Ashton, Conway et al. Nature Neurosci. (2012)

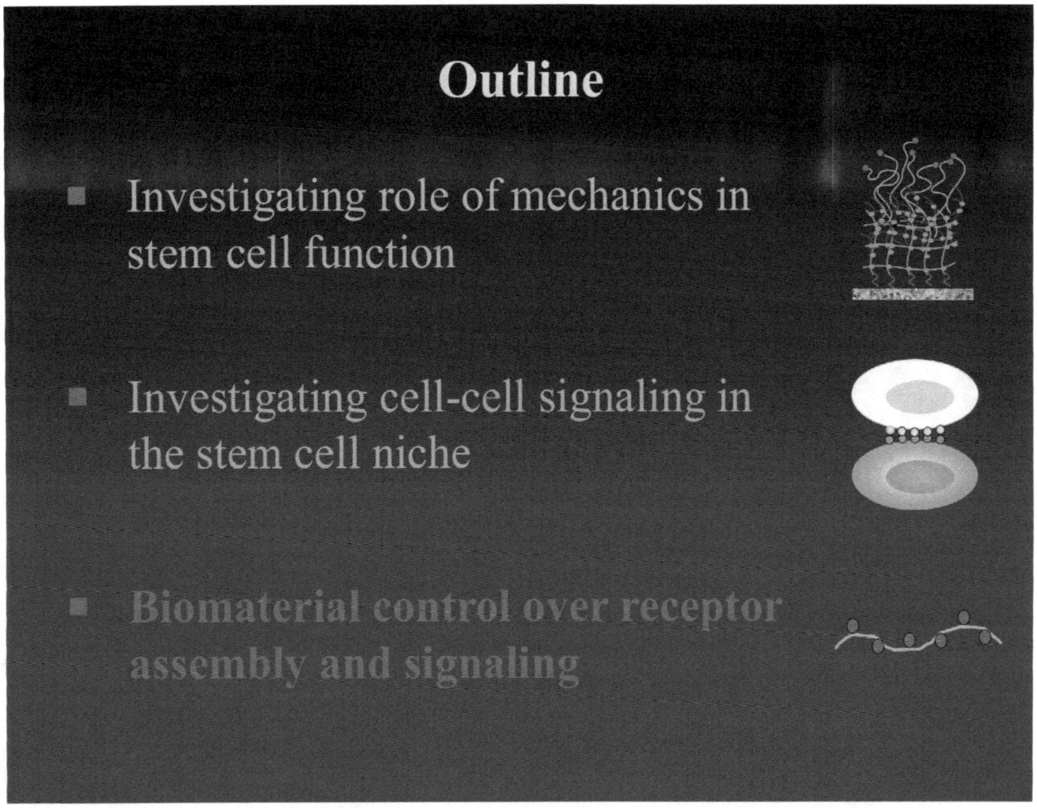
Outline
Investigating role of mechanics in stem cell function
Investigating cell-cell signaling in the stem cell niche
Biomaterial control over receptor assembly and signaling

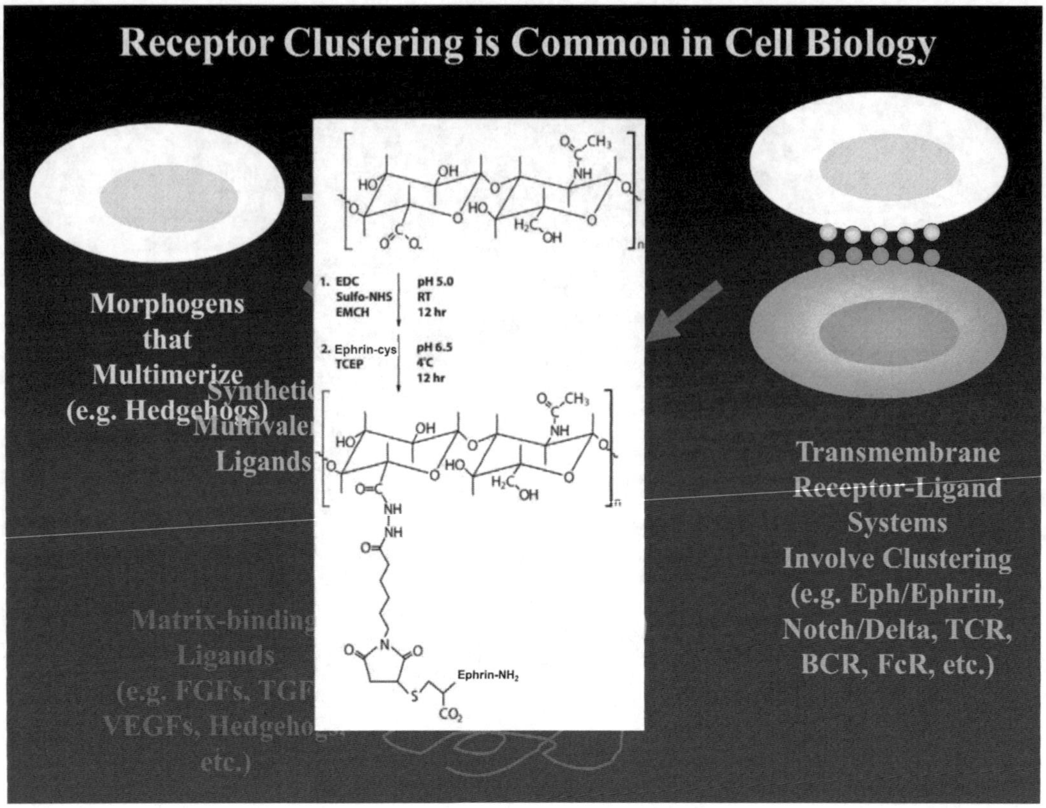
Receptor Clustering is Common in Cell Biology
Morphogens that Multimerize (e.g. Hedgehogs)
1. EDC Sulfo-NHS EMCH
pH 5.0 RT 12 hr
2. Ephrin-cys TCEP
pH 6.5 4°C 12 hr
Ephrin-NH2
Transmembrane Receptor-Ligand Systems Involve Clustering (e.g. Eph/Ephrin, Notch/Delta, TCR, BCR, FcR, etc.)

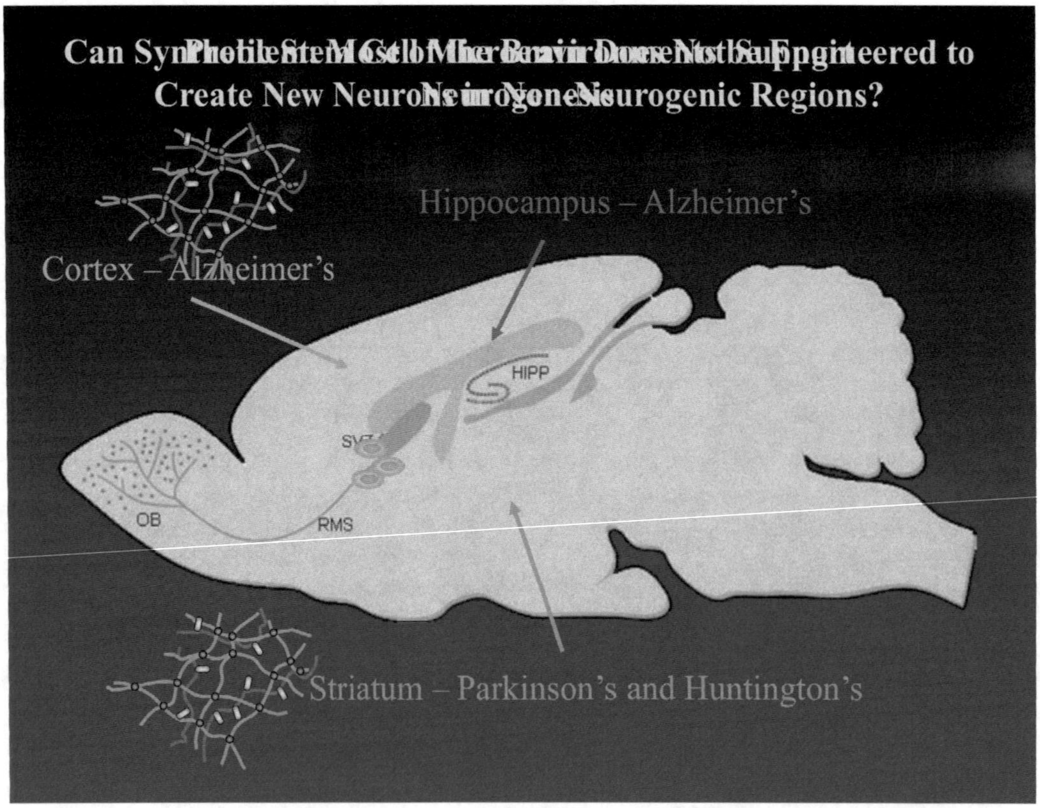
Hippocampus – Alzheimer's
Cortex – Alzheimer's
HIPP
OB
RMS
Striatum – Parkinson's and Huntington's

Components of an Engineered *in Vivo* Stem Cell Microenvironment

- Factor to promote neuronal differentiation: ephrin-B2 (*Nature Neuroscience*, 2012)
- Factor to promote NSC proliferation

Sonic Hedgehog is a NSC Mitogen

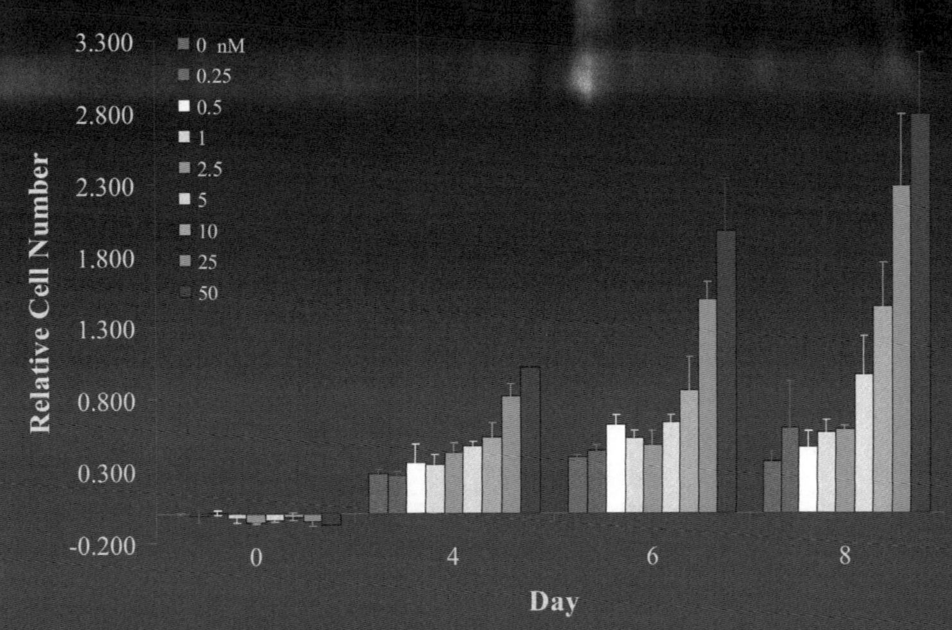

Lai, Kaspar, Gage, Schaffer *Nature Neuroscience* (2003)

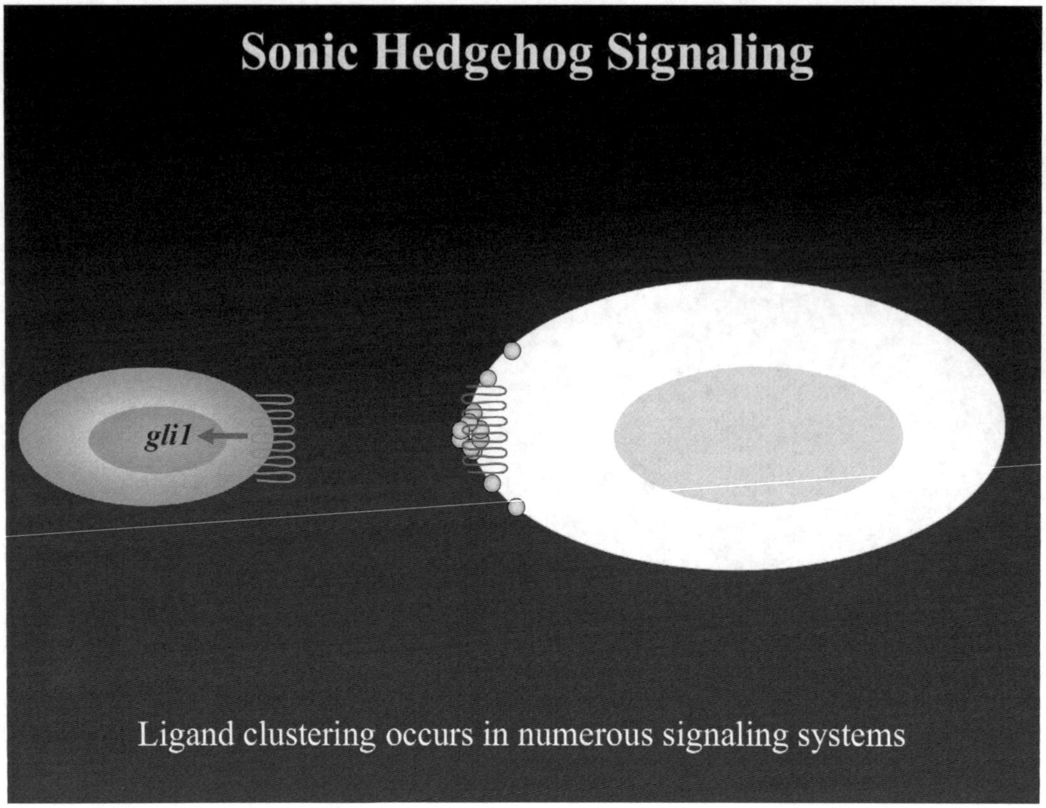
Sonic Hedgehog Signaling
gli1
Ligand clustering occurs in numerous signaling systems

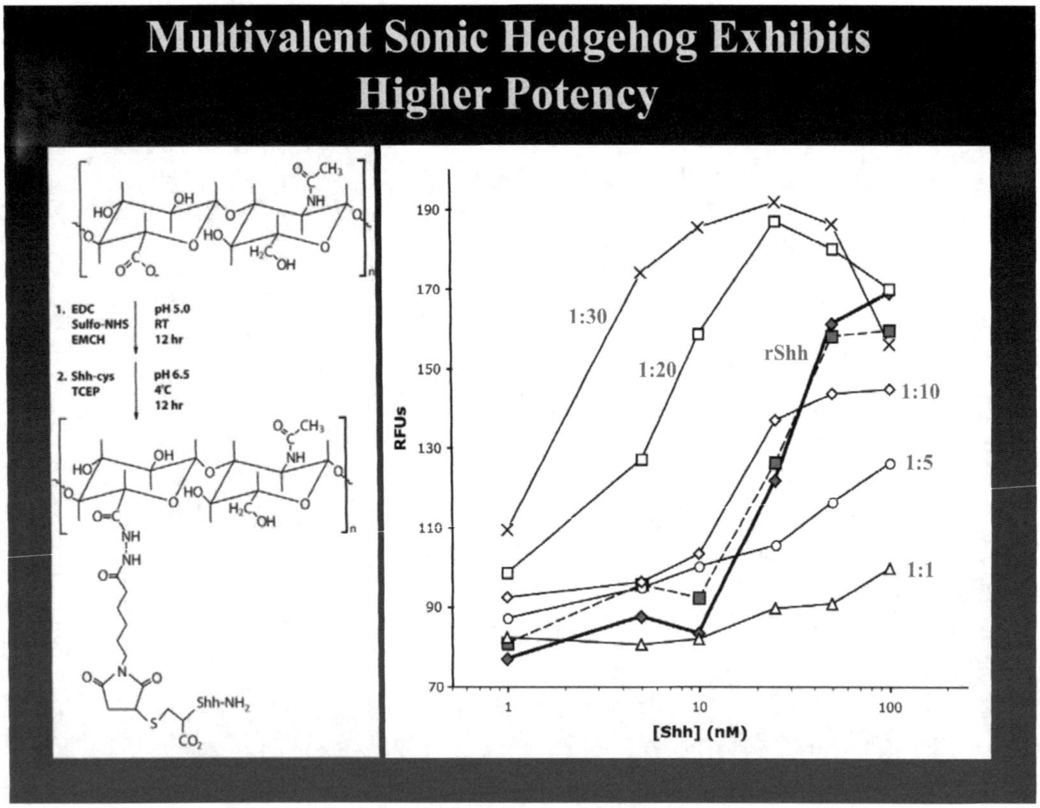
Multivalent Sonic Hedgehog Exhibits Higher Potency
1. EDC Sulfo-NHS EMCH
pH 5.0 RT 12 hr
2. Shh-cys TCEP
pH 6.5 4°C 12 hr
Shh-NH2
RFUs
[Shh] (nM)
1:30
1:20
rShh
1:10
1:5
1:1

Components of an Synthetic *In Vivo* Stem Cell Microenvironment

- Factor to promote neuronal differentiation: ephrin-B2 (*Nature Neuroscience*, 2012)
- Factor to promote NSC proliferation: Sonic hedgehog (*Nature Neuroscience*, 2003)
- Factor to promote NSC migration: SDF-1

Summary

- Must better understand and engineer microenvironmental signaling mechanisms, both biochemical and biophysical, to control stem cell function.
- Ephrins regulate neuronal differentiation of adult neural stem cells and human pluripotent stem cells.
- Biomimetic materials to spatially organize Eph receptors can potently activate signaling and regulate cell fate in vitro and in vivo.
- Engineered biomaterials offer the potential to create synthetic niches that recruit and regulate stem cells *in vivo*.

Schaffer Group

Melissa Bartel
Lukasz Bugaj
Leah Byrne, Ph.D.
Sisi Chen, Ph.D.
Anthony Conway
Tim Day
Siddharth Dey, Ph.D.
Barbara Ekerdt
Benjamin Epstein
Jonathan Foley
Ashley Fritz
Bum-Yeol Hwang, Ph.D.
Leo Lei, Ph.D.
Prajit Limsirichai
David Ojala
Anusuya Ramasubramaniam

Wanichaya Ramey
Sebastian Rammensee, Ph.D.
Jorge Santiago-Ortiz
Dawn Spelke
Marjan Varedi, Ph.D.
Tandis Vazin, Ph.D.
John Weinstein
Hanadie Yousef

<u>Contributing Alums</u>
Randy Ashton
Albert Keung
Kris Saha

<u>Collaborators</u>
Kate Excoffon (WSU)
Kevin Healy
Kathy High (CHOP)
John Flannery
Ravi Kane
Sanjay Kumar
Paul McCray (UI)
Bill Merigan (UR)

NIH, CIRM, NSF, DOE, Foundation Fighting Blindness, Project ALS, ALS Association, Alzheimer's Association, Cystic Fibrosis Foundation, American Parkinson's Disease Association, NSF CAREER Award, ONR Young Investigator Award, Whitaker Foundation,, Hellman Family, Merck & Co., Inc., UC Davis CRCC, UC Discovery, Bio-Rad, Geron, CDMRP TSC

Schaffer Group

Melissa Bartel
Lukasz Bugaj
Leah Byrne, Ph.D.
Sisi Chen, Ph.D.
Anthony Conway
Tim Day
Siddharth Dey, Ph.D.
Barbara Ekerdt
Benjamin Epstein
Jonathan Foley
Ashley Fritz
Bum-Yeol Hwang, Ph.D.
Leo Lei, Ph.D.
Prajit Limsirichai
David Ojala
Anusuya Ramasubramaniam

Wanichaya Ramey
Sebastian Rammensee, Ph.D.
Jorge Santiago-Ortiz
Dawn Spelke
Marjan Varedi, Ph.D.
Tandis Vazin, Ph.D.
John Weinstein
Hanadie Yousef

<u>Contributing Alums</u>
Randy Ashton
Albert Keung
Kris Saha

<u>Collaborators</u>
Kate Excoffon (WSU)
Kevin Healy
Kathy High (CHOP)
John Flannery
Ravi Kane
Sanjay Kumar
Paul McCray (UI)
Bill Merigan (UR)

NIH, CIRM, NSF, DOE, Foundation Fighting Blindness, Project ALS, ALS Association, Alzheimer's Association, Cystic Fibrosis Foundation, American Parkinson's Disease Association, NSF CAREER Award, ONR Young Investigator Award, Whitaker Foundation,, Hellman Family, Merck & Co., Inc., UC Davis CRCC, UC Discovery, Bio-Rad, Geron, CDMRP TSC

Schaffer Group

Melissa Bartel
Lukasz Bugaj
Leah Byrne, Ph.D.
Sisi Chen, Ph.D.
Anthony Conway
Tim Day
Siddharth Dey, Ph.D.
Barbara Ekerdt
Benjamin Epstein
Jonathan Foley
Ashley Fritz
Bum-Yeol Hwang, Ph.D.
Leo Lei, Ph.D.
Prajit Limsirichai
David Ojala
Anusuya Ramasubramaniam

Wanichaya Ramey
Sebastian Rammensee, Ph.D.
Jorge Santiago-Ortiz
Dawn Spelke
Marjan Varedi, Ph.D.
Tandis Vazin, Ph.D.
John Weinstein
Hanadie Yousef

Contributing Alums
Randy Ashton
Albert Keung
Kris Saha

Collaborators
Kate Excoffon (WSU)
Kevin Healy
Kathy High (CHOP)
John Flannery
Ravi Kane
Sanjay Kumar
Paul McCray (UI)
Bill Merigan (UR)

NIH, CIRM, NSF, DOE, Foundation Fighting Blindness, Project ALS, ALS Association, Alzheimer's Association, Cystic Fibrosis Foundation, American Parkinson's Disease Association, NSF CAREER Award, ONR Young Investigator Award, Whitaker Foundation,, Hellman Family, Merck & Co., Inc., UC Davis CRCC, UC Discovery, Bio-Rad, Geron, CDMRP TSC

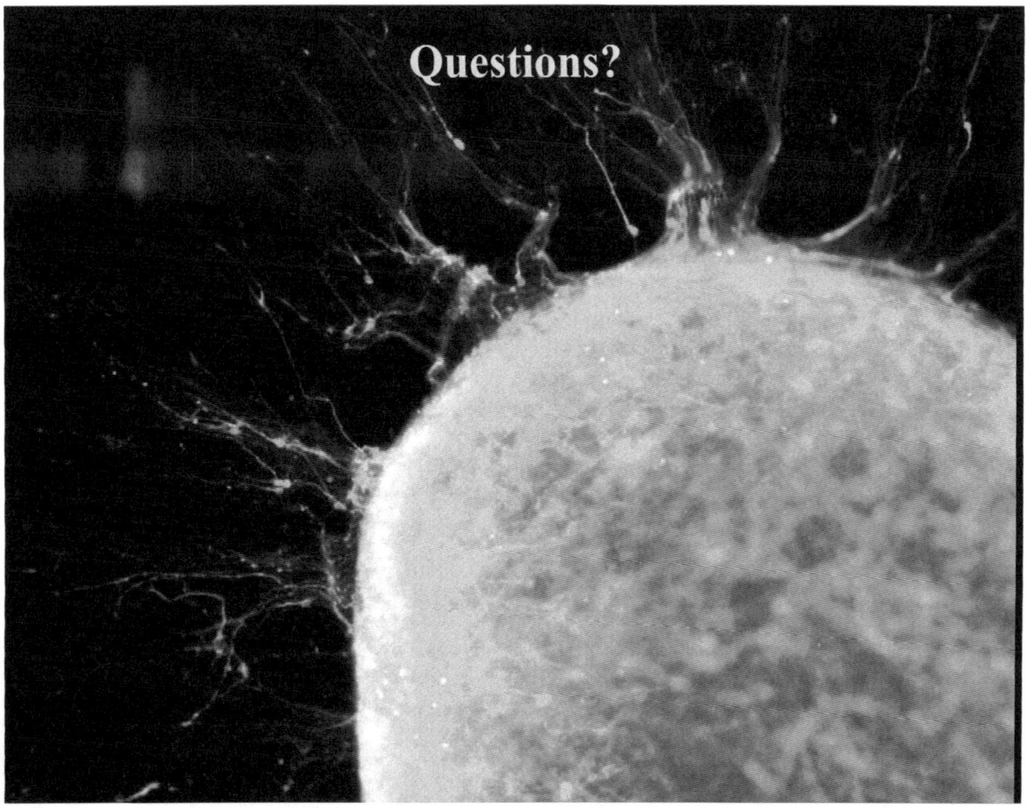

AUTHOR INDEX